Bowtie Methodology

Bowtie Methodology

A Guide for Practitioners

Sasho Andonov

CRC Press is an imprint of the
Taylor & Francis Group, an **informa** business

CRC Press
Taylor & Francis Group
6000 Broken Sound Parkway NW, Suite 300
Boca Raton, FL 33487-2742

Printed on acid-free paper

International Standard Book Number-13: 978-1-138-07997-7 (Hardback)
International Standard Book Number-13: 978-1-138-06705-9 (Paperback)

Library of Congress Cataloging-in-Publication Data

Names: Andonov, Sasho, author.
Title: Bowtie methodology : a guide for practitioners / by Sasho Andonov.
Description: Boca Raton : CRC Press, 2018. | Includes bibliographical references and index.
Identifiers: LCCN 2017018288 | ISBN 9781138067059 (pbk. : alk. paper) | ISBN 9781138079977 (hardback : alk. paper) | ISBN 9781315158853 (ebook : alk. paper)
Subjects: LCSH: Industrial safety--Methodology.
Classification: LCC T55 .A495 2018 | DDC 658.4/08--dc23
LC record available at https://lccn.loc.gov/2017018288

Visit the Taylor & Francis Web site at
http://www.taylorandfrancis.com

and the CRC Press Web site at
http://www.crcpress.com

Contents

Preface

The idea of writing this book came to me after publishing my previous book *Quality-I is Safety-II: The Integration of Two Management Systems*. While writing that book, I wrote about the misunderstandings concerning the quality and safety and in that book I critically assessed the "new star in the safety sky" named as Safety-II. Safety-II is bringing a new method for safety analysis called Functional Resonance Analysis Method (FRAM) and I was wondering, why we need new methods when the older ones are already unknown to the majority of the quality and safety engineers. Looking into the literature, I came to a conclusion that I could not expect that my favorite methodology (Bowtie*) will be used because there are no proper sources of information of when and how to use it. So, I decided to summarize my theoretical and practical knowledge and experience in a book, which will be dedicated to practitioners, and not to scientists.

My present knowledge and experience are very much in favor of the Bowtie methodology (BM) as the best methodology to deal with safety or quality assessment. During the writing of this book, I received a couple of comments that using the term "the best methodology" might be a really strong expression, but my question is: Do you know a better one? No one responded... And this question is still valid: Are you familiar with a method or methodology, which provides more reliable information about safety and quality analysis than the BM? And I am not discussing about a strictly descriptive methodology, but the one which requires the knowledge of something called "science."

Bowtie, through two assessment methods, Fault Tree Analysis (FTA) and Event Tree Analysis (ETA), has a particular level of mathematics built into it, which is usually unfamiliar to engineers in the industry who are dealing with manufacturing or operational processes in their companies. These mathematics are probability, statistics, and Boolean algebra. Although Boolean algebra has a great use in software† for operational systems in computers and processors, most of the electronics' engineers are not so good at it.

In general, I decided to write this book with the intention to provide safety and quality managers with a source from which they can study the BM, and to give them a chance to implement it in their everyday activities.

---

* Bowtie, Bow tie, Bow-tie or Bow-Tie? This is a question that has been bothering me for a while... In the Internet you can find all of them. I spoke to a few English teachers in the Military Technological College and the outcome of this "investigation" was: You can use all of them, but Bowtie and Bow tie are mostly correct. So, I decided to use Bowtie!

† There is a tiny connection between the Boolean algebra and the engineering of computers in some programming languages such as C and Pascal.

But let us clarify something first…

There is already a method under the name BM, which is extensively used in the oil and petroleum industry. The Dutch company Shell is using this method in their everyday operations and it is supported by a software package called THESIS. Since 2010, Petrobras (Brazilian oil company) is also dedicated to using the same BM in their exploration and production facilities, but with the intention to protect the company from incidents and accidents.

This is not the same BM, which I will be speaking about in this book!

This BM (for oil and petroleum industry!) will not be a part of this book and I will not mention anything else in this book regarding this method with the only exception being something minor in section 1.2! In this book, I am presenting the methodology BM, which is made up from FTA and ETA. This methodology is a real representation of what is known as a probabilistic safety assessment.

In aviation, the Eurocontrol–Institute Of Air Navigation Services (IANS) in Luxembourg was using the BM in some of their courses. Federal Aviation Administration (FAA) and Airservices (Australia) are also utilizing it. The Civil Aviation Authority of Singapore (CAAS) has issued ATSIC (Air Traffic Services Information Circular) on January 30, 2009, which shows the use of the BM in hazard identification and risk management. Keep in mind that there are other subjects in aviation, which are also using the BM.

Other areas that use the BM are health care (medicine), pharmacy, transport, nuclear, and chemical industry.

Bowtie as a methodology can be used in many areas, but I will stick to its use in the practical assessment of quality and safety in processes under supervision of quality management system (QMS) and safety management system (SMS). Although this book is based on science, it is not for a scientific purpose. It is also not for dummies, because dummies should not get involved in quality and safety at all. This is a book for quality and safety managers in the industry. It can be used as a textbook in universities where reliability, safety, or quality is taught, but I will let the lecturers decide on that part…

There are variations of FTA and ETA, which are more scientific and more accurate (Kinetic Tree Theory, Binary Decision Diagrams, etc.), but I found them nonpragmatic for practical use in the industry and because of that they will just be mentioned and will not be considered here. Here, I will present a methodology, which I will use to draw the best from all of the available information and all the situations that I encounter in my everyday life, but from an engineering point of view. Someone else may have a different opinion…

This book is strongly dedicated to the practical implementation of the BM in industries where QMS and SMS are implemented. Keeping in mind that my background is aviation, I will mostly use examples from this area assuming that if these examples satisfy the aviation safety and quality criteria, it will also satisfy other criteria in other industries. Having in mind that generally safety is more critical than the quality, in this book I will discuss about safety, but everything discussed here will also apply to quality, maybe with some minor changes.

Why I do believe that this book will be different from the already available books on the market...?

First, I could not find any other book, which is dedicated only to BM. Most of the books that deal with probabilistic safety management mention about the BM, but that is not enough for using it. So, what I am trying with this book is to provide enough material for practical use of the BM in the industry.

Second, I believe that I can provide a simpler and more "user-friendly" explanation to this methodology, which will make it more understandable to the people who need to use it. In the history of my teaching activities, which spread from the time when I was just a simple student in my gymnasium*, I had the opportunity to teach others. As a good student I was asked by the parents of my friends to help them understand mathematics, physics, and chemistry. Even though it did not contribute much to my pocket money incomes, I am (and was) very proud of the results, which I have achieved in this field. The reason for my success was my different approach when explaining about the material and I do believe that this approach still give great results during my teachings in Military Technological College, Muscat, Oman.

Third, this book provides a quality material to deal with risk management from beginning to end in a quantitative and qualitative manner.† It means it starts with the identification of hazards and concludes with the scenarios for elimination or mitigation of the risks, which makes this book holistic. In addition, this book is unique due to the fact that there are plenty of books, which are dealing with FTA and ETA in the market, but there is none that deals with the Bowtie methodology as a whole.

In this book, I am using a lot of examples. Please understand that those examples are simplified with the intention to show how the BM should be used. I will not go too much into detail, because it is not my intention to produce a valid safety case for a particular operation but to explain how the BM is working in different situations. On account of this, some people may not be satisfied with the specifications given in this book.

I really do believe that this is quite a different book for quality and safety practitioners and I hope that you will find it useful...

Sasho Andonov
Skopje, Macedonia

* Gymnasium is a form of high (Secondary) school in the countries of former Yugoslavia, which provides general education.

† Qualitative manner means things will be described with words (descriptive) and quantitative manner is expressing it by the numbers (as a result of measurements or calculations). Keep in mind that it is not always possible to express quality using numbers...

About the Author

Sasho Andonov is a graduate engineer in electronics and telecommunications and has master's degree in metrology and quality management; he obtained both these degrees from the Ss. Cyril and Methodius University of Skopje, University in Skopje, Republic of Macedonia. He has a total experience of 27 years, of which 21 years of experience is in aviation working group in Macedonia, India, Papua New Guinea, and Oman. He has been a member of few International Civil Aviation Organization (ICAO) and Eurocontrol working groups and he has presented his papers in many conferences and symposiums. For the past 13 years, he is dealing with quality and safety management in industry and aviation and he has published a book *Quality-I is Safety-II: The Integration of Two Management Systems.*

Acronyms and Abbreviations

ALARP	as low as reasonably practicable
AOT	actual operating time
ATCo	air traffic controller
BDD	binary decision diagrams
BDMP	Boolean logic Driven Markov Processes
BM	Bowtie methodology
CAAS	Civil Aviation Authority of Singapore
CBA	cost benefit analysis
CEA	cost effectiveness analysis
CET	containment event tree
CFT	component fault tree
CoS	continuity of service
CRT	cathode ray tube
ETA	event tree analysis
EU	European Union
EUROCAE	European Organization for Civil Aviation Equipment
FAA	Federal Aviation Administration (USA aviation regulatory body)
FFA	functional fault analysis
FMEA	failure mode and effect analysis
FRAM	functional resonance analysis method
FT	fault tree
FTA	fault tree analysis
HiP HOPS	hierarchically performed hazard origin and propagation studies
IANS	Institute of Air Navigation Services (training center of EUROCONTROL)
ICAO	International Civil Aviation Organization
IEC	International Electrotechnical Commission
IEEE	Institute of Electrical and Electronic Engineers
ILS	Instrumental Landing System
LOPA	level of protection analysis
LRU	line replaceable unit
MCI	methyl (Iso) cyanate
MCS	minimum cut set
MIL HDBK	military handbook
MSA	measurement system analysis
MTBF	mean time between failures (used for repairable items)
MTBO	mean time between outages
MTTF	mean time to failure (used for non-repairable items)

MTTR	mean time to repair
NFPA	National Fire Protection Association
NOT	non-operating time
OHS	occupational health and safety
QA	quality assessment
QC	quality control
QM	quality manager
QMS	quality management system
RTCA	Radio Technical Commission for Aeronautics
SAR	search and rescue
SEFT	state/event fault tree
SFT	sub-fault tree
SFTA	software fault tree analysis
SiS	signal in space
SM	safety manager
SMS	safety management system
SPC	statistical process control
SPF	single point failure
SS	sample space
SUV	sport utility vehicle

1

*Introduction to Bowtie Methodology**

1.1 Introduction

The Bowtie methodology (BM) is a methodology that specifies two methods for one more general model (theory) for risk analysis.

This general model (theory) states that there is a risk of something to happen and there are previous and later activities to help us analyze and eliminate (mitigate) the risk. Previous activities for risk analysis are methods that are used to find the cause for the event happening (determined by a particular risk), while later activities are used to explain the scenarios of the development of the consequences caused by this event. From this general definition of the model for risk analysis any two methods can be used.

The BM is more specific by using two methods for previous and later analysis of the so-called Main Event that can happen in the system. The first method is Fault Tree Analysis (FTA) and the second one is Event Tree Analysis (ETA). In the BM configuration, FTA is known as Pre-Event Analysis and ETA is known as Post-Event Analysis.

The BM gets its name due to the shape of the diagram which arises when these two methods are connected to the point called Main Event (Figure 1.1). A Main Event can be connected with a system, process, activity, operation, situation, measurement, and so on. That brings us to the main point of the BM: It can be applied to systems, processes, activities, operations, situations, specifications, and so on. Whatever the Main Event is, we must know **where** (in which system, which part, etc.) and **when** this event (operation, activity, etc.) could happen.

* If you go on the internet and search for Bowtie, you will find a plenty of articles and in most of them Bowtie is named as a method. That is wrong because a method is actually a procedure on how to deal with a particular operation (activity) and a methodology is the aggregation of methods. So, knowing that the Bowtie consists of FTA and ETA (which are methods) it is clear that it is in fact a methodology.

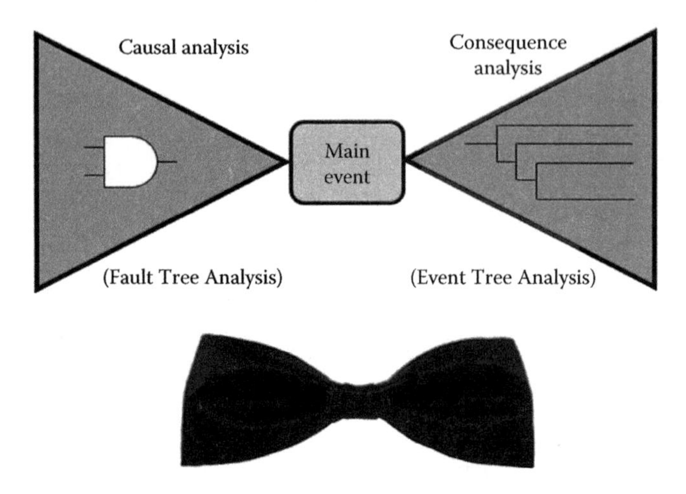

FIGURE 1.1
Bowtie methodology (BM) diagram and ordinary Bowtie. (In some literature you can find the graphical presentation of Bowtie as the Cause-Consequence Diagram (CCD).)

Do not forget that in the beginning, the BM was linked to the reliability of equipment* and it proved itself good in this area. FTA was used for reliability calculations and ETA for measures to improve the reliability. Reliability was expressed as trust in the normal functioning of equipment and later it was spread to Risk Management. This happened especially in risky industries in the times when malfunctioning of equipment was counted as the main reason for incident and accidents.

It is important that both methods (FTA and ETA) can be used independently, but what connects them in the BM is the Main Event. This connection causes dependence between them, because changes in one of them can cause changes in another. Both methods will be described in detail later in this book, but only in the context for their use in the BM.

The main purpose of the BM is to provide accurate and reliable data for the decision-making process. This decision-making process is twofold. First, I use results from FTA to calculate the probability of the Main Event happening and then I use ETA to calculate the success or failure of the particular mitigation for consequences.

The area of use of the BM is wide. It can be used during the design phase, during implementations, system operations, and system modifications, but it provides best results when implemented during the design phase! At the beginning in the design phase, by using FTA (Pre-Event Analysis), I can calculate the frequency (probability) of the failures (Main Event) happening

* Today's equipment is made of hardware and software, and in this book, the term "equipment" means an aggregation of both software and hardware.

for the new product. Later, ETA (Post-Event Analysis) will give me a clear picture about the possibilities and scenarios to contain or mitigate the risk if the failure (Main Event) happens.

Here I should mention a very important point: the BM does not determine the severity of the Main Event. It must be determined in another process by taking into account the nature and environment of the Main Event. Anyway, to establish severity I will use the criteria that deal with the real situation of the systems under analysis. For different industries or even for different systems in the same industries, these criteria may be different.

What is also important that I should mention here is that the BM can also be used for a better understanding of different levels of functioning of processes in our management or production system. So, it is not necessarily connected with problems, but having in mind that the BM is highly descriptive, I can use it for quality control (QC) or quality assurance (QA) to improve our product or services offered.

1.2 BM in Oil and Petroleum Industry

I have already mentioned in Preface that there is another method with the same name (Bowtie) that is already used in the oil and petroleum industry.* I would like to explain here why I do not appreciate it very much.

This is a BM that deals, in the first instance, with hazards (causes) and barriers (controls) to the causes and, in the second instance, with mitigation and recovery actions for consequences. It is a method, not a methodology, and it starts with hazard identification. You may use any tool for hazard identification, and when the hazards are identified, the particular barriers are set to stop these hazards from developing into incidents or accidents. So, this is mostly a prevention method.

This is actually a method that presents the famous ***Swiss cheese*** model (Figure 1.2) of barriers and holes in the barriers. It is mostly used to build barriers for eliminating or mitigating safety hazards and later to check the effectiveness and efficiencies of already implemented barriers.

In Figure 1.3 are presented the differences between the method and the methodology.

This method does not use FTA for Pre-Event Analysis and ETA for Post-Event Analysis to calculate probabilities, which means that this is a method that does not calculate risks. It starts from hazards (not from the Main Event!) and later is upgraded with barriers and particular controls to prevent

* Somewhere I read that this method was proposed by Valerie De Dianous and Cecile Fievez in 2005, but I cannot confirm.

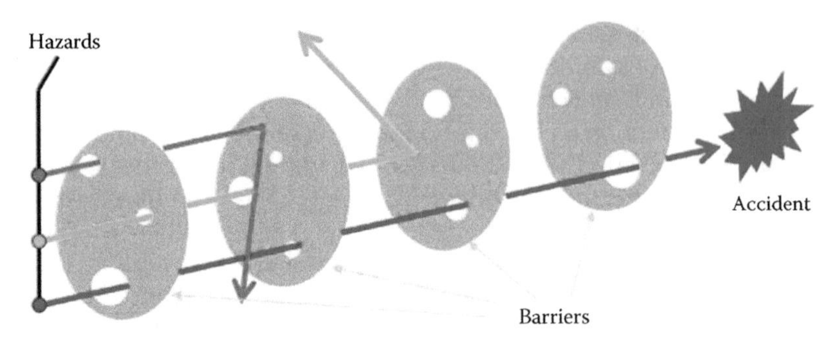

FIGURE 1.2
Cheese model.

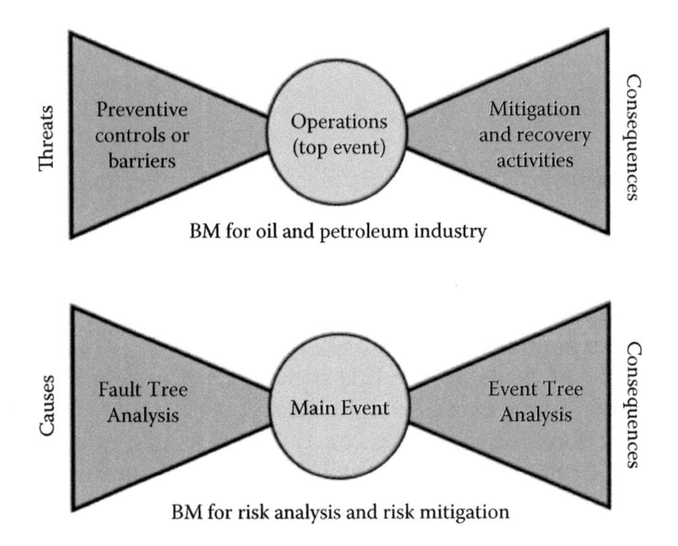

FIGURE 1.3
Comparing the BM in oil and petroleum industry with the BM for risk analysis and risk mitigation.

abnormal operations. After that, under the assumption that abnormal operations happened, there are particular measures for mitigation of the consequences or recovery actions to get back to the normal operations. All this is built under the analyzing concept of layers known as Layers of Protection Analysis (LOPA). It means that all barriers are posted by layers (one layer may contain a few barriers) and this structure provides a better understanding of the safety system dedicated to this operation. For example, Petrobras have nine such layers. Five are used to prevent the event and four are used to mitigate the consequences and provide fast recovery to normal operations. Associated with this method is the particular system for integrity monitoring in real time.

There are a few points that I do not like in this method.

The main point is that this method mixes threats with hazards and risks. For me that is very confusing. Maybe it works for employees, but not for me. Threats and hazards do not differ as much and risk is when the hazard is determined by its frequency (probability) and severity. Actually, as far as I know, threats are synonyms for hazards. This method associates hazards with the Main Event (which is named as the Top Event) and threats are actually *taking the places of hazards in this diagram*. But this is not a big issue if you notice that plenty of regulatory bodies make no difference between these terms. Even in the ISO 31000 standard (Risk Management) you cannot find the word "hazard" at all.

The second point (which I do not like here) is the fact that there is no explanation of how barriers are posted on particular positions. It means that this method does not provide us a real picture of what risks are for each hazard and how the risks progress after the posting of the barriers or preventive controls. To find this you need to use another method, which means that this one offers benefit only to the monitoring personnel who are overseeing normal operations or to regulatory staff who can see how the company is dealing with risks.

The third point is that when abnormal operation happens, there is no knowledge about the probability of success or failure of the mitigation or recovery action. Of course, the reason why these barriers are there is not clear and their quantification should be calculated by another method.

The fourth and maybe the most important point is that there is no information about how barriers, control measures, mitigation, and recovery activities depend on each other and how they influence the other parts of the system.

So this is an excellent descriptive method that can be used only as an explanation of how companies are providing and maintaining the safety of their operations, but its contribution to the safety assessment process is very poor. The main point is that after finishing with the BM for Risk Analysis and Risk Mitigation, you may produce this type of BM just to explain to employees, regulatory bodies, and public how they can handle safety in their operations. But I do not think that using it as a tool for the analysis and management of a risk can add any value.

In this book only the Bowtie methodology (not the method!) will be explained.

1.3 Context of Investigations in Science and Industry

I would like to mention here a very important point that is connected to our reality. I would like to speak about the context of points in life, science, and industry.

Let us speak about the context in our lives and let us say I would like to buy car.

There are many particular specifications that I would like my car to satisfy and all of them are dependent on my personality, my incomes, and my requirements. There are plenty of car manufacturers in the world. They differ by the models and classes of cars manufactured. The best ones are Jaguar, Porsche, Ferrari, McLaren, Bugatti, BMW, and so on. If the context of buying my car is to emphasize (or build) my status in the society, I will buy some of these cars.

There are also family cars produced by Opel, Toyota, Nissan, Volkswagen, Hyundai, and so on. Which car I would personally buy depends on the context of my requirements. If I need a car for simply driving in the city, I will buy a small, low-fuel consumption car that will not cost me a lot to buy it and to maintain it in the future. If I am an adventurer and I enjoy nature (sea, mountain, etc.), then I will buy 4 × 4 SUV (sport utility vehicle). If I have some small business in painting or home appliance services, I would probably buy Toyota Hilux or Dodge Ram for putting my tools and materials in the load box behind. I know what is best on the market, but I will personally choose a car that fits my requirements. This is how clever guys live their lives.

Science is dedicated to understanding the functioning of our world, so it applies methods that are accurate and precise as much as it is possible. Sometimes few decimals in calculations make a huge difference. For nuclear physicist, dimensions of electrons, neutrons, and protons are very important to explain the "functioning" of atoms, but for electrical engineers, dimensions are not important because they treat them just as a bunch of moving electrons and ions. So, when scientists produced the cathode ray tube (CRT), they took into consideration everything. But when electronic engineers build equipment with CRT, they do not bother themselves with the dimensions of electrons, even though the flow of electrons is the fundamental principle of the CRT functioning. Obviously, the context of using electrons for nuclear physicists and electrical engineers is different.

The same point generally applies for science and industry too. Although science is the foundation of industry, the context is quite different. For science, quantity is more important, and for industry, quality. Science is looking for exact numbers and industry is putting those numbers in the context of specifications and tolerances. For industry, numbers are used only to see if they fit or do not fit tolerances. And this difference is crucial when using science in industry.

The same points happen with the methods used for measurements or conducting and controlling processes in science and industry. Requirements for measurements in science and industry are based on their context, and laboratories in the industry are quite different than the scientific laboratories. Scientific laboratories are used for measurements in controlled environments where temperature, pressure, and humidity are kept at a particular value. Laboratories in industry are not so stringent in these requirements. They may be, but almost always the industry products will not be used in a controlled environment. In fact, they will be used by humans who do not care about the environment. Cars produced in car factories will be used in Sweden and in Saudi Arabia, where the environment is extremely different. So, even though the car is tested in laboratory, it must be tested and adapted to different environments, which is imperative.

Although science has produced plenty of methods for measurement, monitoring, or controlling the processes in industry, the industry is prone to make compromises due to economic reasons. The reason is simple: The industry is governed by customers, and if the customers are satisfied with the particular accuracy and precision of the products or services which are offered, then the industry is also satisfied.

The regulations in aviation, regarding landing of aircraft, have three classes of aerodrome operations (CAT I, CAT II, and CAT III)* which have quite different specifications and requirements. CAT I is with low requirements regarding the availability, reliability, integrity, and continuity of service; CAT III is with high requirements; and CAT II is somewhere in between. It is clear that CAT III is safest, but there are plenty of aerodromes that are not CAT III certified. There is no economic feasibility to implement the CAT III operations there, simply because the level of traffic is too low. There are also aerodromes that simply cannot fulfill environmental requirements for CAT III operations,† so implementing CAT III operations is not possible at all.

The BM with FTA and ETA actually proves the difference in context between science and industry: There are plenty of methods that give more accurate results for safety analysis compared with these two, but industry does not detect the economic feasibility to implement them. So, I find this difference of context to be very important when dealing with reality, and thus I will use it in this book. Plenty of scholars might not agree with my comments or recommendations here, but I find those comments irrelevant when dealing with industry.

1.4 BM and Software

The BM is fully applicable for both quality and safety assessment of software during its design process. FTA is often used as a method under the name Software Fault Tree Analysis (SFTA) with FAA and NASA. The reason to use SFTA is twofold: to find particular operational conditions if the software

* More explanation for this part will be given in Chapter 8.
† More details can be found in ICAO Annex 14 (Aerodromes) document.

fails or to prove that specific software design is safe (will operate without safety failures). First I suppose that the software has found itself in an event that is determined as an unacceptable risk and, using SFTA, I am going backward to investigate the combination of possible causes for that event to occur. SFTA should help to find the causes for the failure event to occur or to demonstrate that this event is impossible to occur if such conditions are present.

The designing of the software is done in accordance with its use in a particular system, but software is, in general, more complicated than hardware. It means that more attention should be dedicated to designing reliable software. That is the reason why, due to the criticality of software in systems in risky industries, there are particular rules and regulations that need to be followed. International Electro-technical Commission (IEC) has published a document (standard) IEC 61508 named "Functional safety of electrical/electronic/programmable electronic safety-related systems," consisting of seven parts. This standard is not part of occupational health and safety (OHS), but it is dedicated to the safety of the products* in all industries and it consists of rules on how to build safety-related systems. Two of the seven parts of this standard are dedicated to software.

So, there are already regulations dedicated to designing software and they explain how software should achieve acceptable integrity when produced for use in safety systems. But before I start with detailed explanations about BM for software analysis, I would like to speak about something else. Actually, I would like to emphasize a very important point that is often forgotten by quality and safety managers during their assessments: Analysis of the software that is already implemented in the systems.

Today the industry is based on technology which is using both hardware and software, so each piece of the equipment should be assessed as an aggregate of both. Knowing that, keep in mind that when I mention "equipment" in this book, I am talking about equipment that consists of both hardware and software. And when I mention "software" I am not talking about software for safety critical systems, but about software that is part of the normal equipment. In other words, software that is not part of the IEC 61508 standard, but one that must not be forgotten when using the BM in already installed equipment.

The main difference between hardware and software, from the quality or safety point of view, is that software does not wear out. It means if something is wrong with the software, it is "produced" during the design process. In addition, the maintenance of software is not done very often, and it is done by different kinds of "maintenance people" compared with hardware maintenance.

* Products here are not only retail products, but also equipment sold as products in industry (aircraft, nuclear reactors, signalization systems for railway, navigation, communication and surveillance systems for aviation, etc.).

So, the role of software in normal or abnormal functioning of equipment is very often neglected. Due to the not wearing out of software, it shows extremely good reliability compared to that of hardware. For example, I read somewhere that in the automotive industry less than 0.1% of all safety issues are due to software problems and 99.9% are due to hardware (mechanical or electrical) issues.

I can determine software's reliability using the quality specifications of the software regarding its capability for error proofing, error prevention, fault detection, recovery, and safe uninstallation and reinstallation. Not fulfilling these quality specifications may cause strong safety consequences and that is the reason why during the execution of the BM, ordinary software should be equally important just like everything else that may be part of the assessment.

Anyway, the main point is: Do not forget to assess software together with hardware during BM executions on real systems!!! Even if you build two fault trees (one for hardware and one for software), you must not forget to connect them in accordance with the interface where they meet each other.

1.5 Main Event

Main Event is the central part of the BM.

As I said in the beginning, a Main Event can be connected with other events, processes, activity, operations, situations, failures, successes, measurements, and so on. Whatever the Main Event is, it should be accurately determined and clearly explained in the frame of the system used.

There are two possibilities, you either like or dislike the Main Event. If you like it, you would like to find the ways of how to make it happen very often. But, if you do not like it, you would want to prevent it from happening again. Of course, this "like–dislike" situation is not subjective. It actually depicts normal situation (like) and abnormal situation (dislike) in the context of quality and safety. If I like the event (pleasant event or normal operation), I can mark its outcome as a Success (S) and in such a situation I will use the BM to deal with Success.*

Unfortunately there are some limitations of normal operation and I need to be careful. Humans are well known by the fact that, sometimes, abnormal situation (if not treated for a long time) is accepted as normal. Let me remind you: If sometimes I have some health problem and I do not treat it (especially when the consequences are not so strong), in time I will get used to think about

* See Section 6.5 (Failure and Success).

it as a normal situation. This is pretty much valid for people with chronic diseases, who are not persistent enough to deal with it in a timely manner.

If I do not like the Main Event (in this case it is an adverse event or abnormal operation), I can mark its outcome as a Failure (F). Going further in this direction I can take care of the Failures of operations, activities, and processes because the irregularities (or variables) in them can produce the Main Events. And that is what the BM is mostly used for: providing enough data for decision making about preventing the adverse Main Events (Failures).

Please note that this is a general explanation of Main Event: It can be a Success or Failure. In this book, mostly I will speak about the use of the BM where the Main Event is a Failure.

In that context, as a definition of the Main Event, I can offer you the simplest one that totally fulfills the requirements of this book: Main Event is a nonsuccessful result* (outcome, effect, etc.) of some activity, operation, or process. The important point with the definition of the Main Event is that now I can dedicate myself to the specifics of my system: I need to learn about the systems behavior to understand how it works normally. Later, this will help me to find how it differs from normal operations, and the BM will help me to investigate how and why it happens.

In general, the Main Event can be caused by system or equipment fault, human error, or by organizational error within an operation, activity, or process. Speaking about the Main Event as failure, I must explain that this is the result of the functioning of the operation, activity, or process outside the specified limits. The reason for that should be internal and external interactions with parts inside or outside the system. Internal interactions should not happen if the system is well designed, extensively tested (controlled), and regularly maintained. External interactions can be harmful if the system is not used in conditions that are specified for its operation. So, I can say that organizational or human errors lead to malfunction (abnormal operational situation), which cause the system to fail due to exceeding system specifications.

The Main Event can be chosen from the registered failures or faults or can be anticipated. Designers that design the system should be aware of these failures and faults, so they can put barriers and controls to prevent them from happening. In these cases, they use the BM† to anticipate the possible scenarios (mechanisms, modes, and effects) how the system might fail.

Now I would like to speak about prioritizing Main Events in industry. If I am dealing with quality or safety management in an industrial company, I will need to produce a list of all Main Events (Failures) that threaten my normal operations. When I have a list I can use it to notice that not all of them happen with the same frequency and the same consequences.

---

* By definition failure is a nonsuccessful result or outcome of something.

† This happens with the BM method used in oil and petrochemical industry!

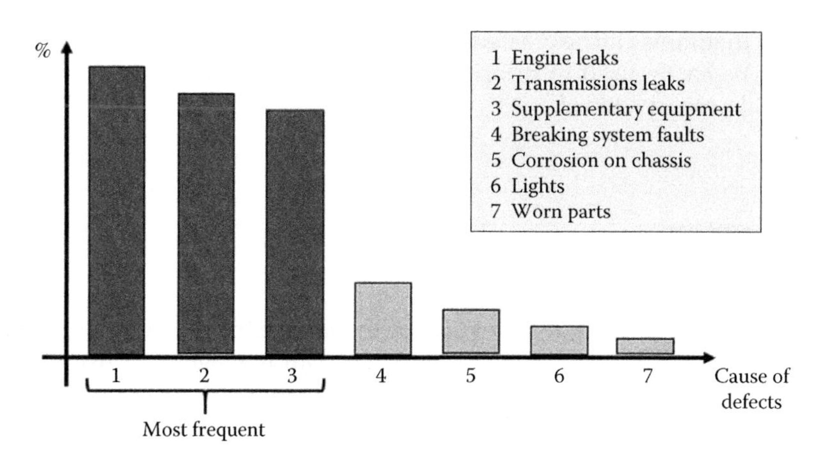

FIGURE 1.4
Pareto diagram.

Of course, a good quality or safety manager will conduct BM to all of these Main Events, but the ones that have severity of accidents (catastrophic events) shall be prioritized. So, there is a need to apply BM first to the events that severity will be accident and incident, even though they happen with low frequency. Later, I will investigate the events with higher frequency (those that happen more often). Having a particular recording and analysis can provide data that can be used with the Pareto diagram to establish the frequency of Main Events.

Vilfredo Pareto was an Italian sociologist and economist who was investigating the distribution of richness and poverty in Europe at the end of the nineteenth and the beginning of the twentieth century. During his analysis, he noticed that 20% of the people actually own 80% of the wealth in Europe.* Later it was discovered that this principle does not apply only to economy, but also to industry. So, the Pareto diagram started to be used in industry where almost 80% of the failures in processes are caused by only 20% of the process variables. It means that controlling 20% of these process variables you will achieve 80% effectiveness in process control.

Figure 1.4 shows a Pareto diagram† where rankings of car defects (by their frequency) are given. You can notice that three of them are marked with the biggest number, which means they happen very often. So, car manufacturers should focus their efforts to improve quality of these three areas and later try to improve others. Solving causes for these three defects will improve image of the cars produced by the manufacturers.

* Pareto diagram is also known as the 80/20 rule.

† The diagram is arbitrarily produced by me and it is used just as an example (which means it cannot be used for other purposes).

Pareto diagrams can be created in Excel, so it is a beautiful and simple tool that can be easily used to prioritize Main Events. It already has its use in industrial process control.

1.6 Risk Management

Risk Management is about how I choose to manage the risk. I need to determine its value (Risk Assessment) and later I can manage it by using some actions for elimination or mitigation (Risk Mitigation).

The BM is a very valuable methodology in that context, because the combination of the FTA and the ETA is serving as a complete tool to deal with the Risk Management. Although the Main Event can be different (event, process, operation, activity, etc.) using BM for the analysis of the Main Event means that I am executing the BM mostly for probability analysis. Saying that means that the BM will provide us with data about the probability of this particular Main Event happening and data for probability that the particular scenarios for consequences and mitigation will result in success or failure. Having in mind that FTA is used for Risk Assessment and ETA is used in Risk Mitigation by using the BM, I can execute a complete Risk Management from the beginning to the end.

Risk Management is about managing risks and it is used in different industries. It is important to clarify some terms connected with Risk Management because even in the official documents there are many differences in this area. The biggest misunderstanding is mixing the terms of hazard and risk. Even the International Standard Organization (ISO), in its published standard ISO 31000 (Risk Management—Principles and guidelines), does not state the difference between hazard and risk. If you open electronic version of this document (word, pdf, etc.) and execute FIND for the word "hazard," it will appear 0 findings.

In general, a hazard is a situation with a capability to harm humans and assets.* This harm can be complete (death of humans or destruction of assets) or partial (injuries for humans and harm of assets that can be fixed with the engagement of humans and money). When I calculate a "capability of hazard to produce harm" I am actually calculating how strong the consequences† of the hazards can be and how often (with which frequency) could this hazard happen. It means I am determining the risk of this situation to happen and

* For the purpose of this course I will define assets as resources of everything that has any kind of value to us: personnel, environment, nature, products, services, equipment, reputation, and so on.

† Consequences, effects, and outcomes have the same meaning in this book!

the consequences of it if it happens. So, I can define risk as a quantified hazard in the manner of frequency* and level of damage done.

If I would like to be scientifically correct, I should use the term "probability" instead of "frequency," but not always "probability" can be used. The reason for this is that the mathematical term "probability" is based on a huge number of experiments and not always there is enough quantity of data to calculate the probability of some event in reality. Having reality in mind, I will sometimes use "frequency" instead of "probability" in this book. It means that if I calculate the probability, it will produce data with more integrity than by calculating the frequency. The difference between frequency and probability is something that affects the accuracy and integrity of the BM. Also, the probability (as a mathematical term) is quantity (numbers) and frequency can be described as both quantity (with numbers) and quality (in words: often, not very often, sometimes, rare, never, etc.). Of course, descriptive words for frequency are used where events happen just a few times, so any calculations regarding the frequency are vague.

Speaking about the level of consequences I may say that there is no quantitative measurement for them. Usually it is descriptive (by words) and different risk industries have different criteria for qualitative measurements. These measurements are usually given in the form of a table and/or matrix. Just as an example, I am offering to you Tables 1.1 and 1.2 where the aviation criteria for risk assessment mentioned in ICAO DOC 9859† are presented. Please note that Tables 1.1 and 1.2 are a way for risk determination in aviation and they include risks expressed by likelihood (frequency) and consequences.

TABLE 1.1

Risk Assessment Matrix

Risk Probability	Risk Severity				
	Catastrophic A	Hazardous B	Major C	Minor D	Negligible E
Frequent 5	5A	5B	5C	5D	5E
Occasional 4	4A	4B	4C	4D	4E
Remote 3	3A	3B	3C	3D	3E
Improbable 2	2A	2B	2C	2D	2E
Extremely improbable 1	1A	1B	1C	1D	1E

(Continued)

* In some literature you can find the term "likelihood" that is equivalent to the term "frequency" used in this book.

† ICAO Doc 9859: "Safety Management Manual"; Third Edition, issued by International Civil Aviation Organization in 2013.

TABLE 1.1 (*Continued*)

Risk Assessment Matrix

Likelihood	Meaning	Value
Frequent	Likely to occur many times (has occurred frequently)	5
Occasional	Likely to occur sometimes (has occurred infrequently)	4
Remote	Unlikely to occur, but possible (has occurred rarely)	3
Improbable	Very unlikely to occur (not known to have occurred)	2
Extremely improbable	Almost inconceivable that the event will occur	1

Severity	Meaning	Value
Catastrophic	Equipment destroyed, multiple deaths	A
Hazardous	A large reduction in safety margins, physical distress, or workload such that the operators cannot be relied upon to perform their tasks accurately or completely, serious injury, major equipment damage	B
Major	A significant reduction in safety margins, a reduction in the ability of the operators to cope with adverse operating conditions as a result of an increase in workload or as result of condition impairing their efficiency, serious incident, injury to persons	C
Minor	Nuisance, operating limitations, use of emergency procedures, minor incident	D
Negligible	Few consequences	E

TABLE 1.2

Risk Tolerability Matrix with Explanations

Assessed Risk Index	Suggested Criteria
5A, 5B, 5C, 4A, 4B, 3A	Unacceptable under the existing circumstances.
5E, 5D, 4C, 4D, 4E, 3B, 3C, 3D, 2A, 2B, 2C, 1A	Acceptable based on risk mitigation. It may require management decision.
3E, 2D, 2E, 1B, 1C, 1D, 1A	Acceptable.

Red	*High (intolerable) risk*: Immediate action required for treating or avoiding risk! Cease or cut back operation promptly if necessary! Perform priority risk mitigation to ensure that additional or enhanced preventive controls are put in place to bring down the risk index to the MEDIUM or LOW OR NO RISK range.
Yellow	*Medium risk*: Shall be treated immediately for risk mitigation. Schedule for the performance of safety assessment and mitigation to bring down the risk index to the LOW OR NO RISK range if viable.
Green	*Low or no risk*: Acceptable as it is. No further risk mitigation required.

The colors in Table 1.1 represent different categories of the acceptability of the risk and they are explained in Table 1.2.[*] Using the qualitative calculations of probabilities from the BM and determining particular consequences I can determine the risk. Comparing the results with the matrix in Table 1.2, I am able to make decisions about the risk: May I live with it or not. If I cannot live with that, it means that I should engage myself to execute further action in the form of elimination or mitigation.

1.7 ALARP

ALARP is acronym for "As Low As Reasonably Practicable" and this is a part when risk management and economy meet each other. Dealing with risk with a successful way has sense only if the price for that is not so high. It means that the benefit of the elimination or mitigation of the risk should be bigger than the price paid for it and this cannot be always achieved. That is the reason why you should deal with a decision-making concept for Risk Mitigation called ALARP.

ALARP in economy is based on the Cost Benefit Analysis (CBA), which is an economic term to compare the benefits and losses of a particular business operation. The CBA is used for particular decision making in economy, and even though there are clear mathematical expressions connected with the CBA, its implementation to Risk Management should not be based only on numbers. The CBA uses a particular set of data for calculations and these data are very often based on "business instincts" that make the CBA highly subjective. The application of the CBA in Risk Management is based on numbers that are usually the probabilities of "bad things" happening and costs of elimination, mitigation, and recovery of normal operations. Costs are easy to calculate and to trust, but probabilities are not so certain. So, overall safety and operational experience associated with the safety intuition should also have a particular role in the determination of the benefits compared to the price of the mitigation.

Speaking about numbers, in economy business is sustainable if costs are less than half of the benefit, but I am pretty much sure that this is not applicable for Risk Management. The reasons for that are the human, social, and environmental aspects of the risk, which are pretty complex for cost calculations. There is no price expressed in money for human lives and environmental damages! So for Risk Management it is better to use the Cost Effectiveness Analysis (CEA). This term comes (same as the CBA) from

[*] ICAO Doc 9859: "Safety Management Manual"; Third Edition, issued by International Civil Aviation Organization in 2013.

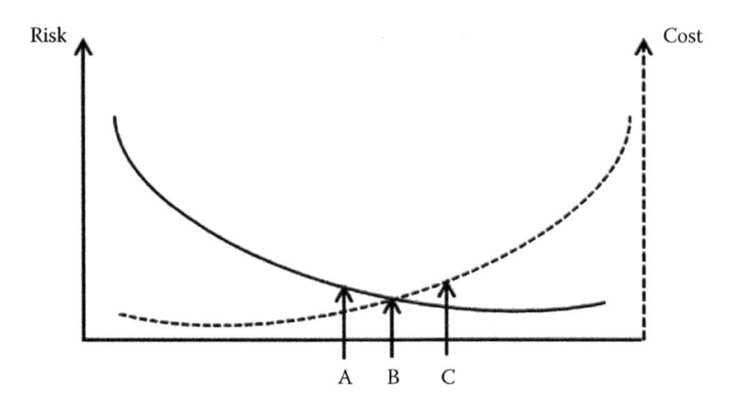

FIGURE 1.5
Choosing ALARP.

economy, but instead the cost (price, money, etc.) I am using different terms expressed by the percentage of the effect achieved. Using this in reality means to be able to calculate the effect of protection against the damage that can be produced by a particular risk consequence (if the Main Event happens). The CEA does not use money to calculate the benefits; instead it uses physical units (decreased area of pollution, lives saved, assets protected, etc.).

In Figure 1.5 is presented a mathematical way of choosing ALARP by CEA* for Risk Management. This is not much different than in economics, but in economy I can apply decision making with a particular gambling (calculated risk). In Risk Management I must apply whatever action is necessary to eliminate and mitigate the risk, which can cause loss of human lives, catastrophic environmental damage, or total destruction of assets.

The full line represents the risk and the dashed line the costs. There are three different choices for ALARP: A, B, and C. CEA would recommend choosing point B where there is optimal balance between the costs and risk. Some of the managers would think that the costs are still too high and they can choose point A, putting less money in mitigating the risk. If I need to choose, I will choose point C! I would simply put more money to mitigate the risks having in mind that having smaller risk gives me some "back up" space in case I have miscalculated something.

There are also other points to consider regarding the ALARP and CEA in Risk Management. Usually two possibilities are present: to eliminate and to mitigate the risks. An important point to consider is that in safety I should always strive to eliminate the risk. So, even though the calculated CEA for both elimination and mitigation showed that mitigation is cheaper than

* If I can express damage in terms of money, then the CBA can also be used!

elimination, elimination shall be our first choice. The reason for that is very simple: I will never know all possible scenarios how the particular hazard will turn into a risk and I will not consider all scenarios that the risk will make damage. So by eliminating the risk, I am eliminating consequences: I am on the safe side!

My recommendation is to use the CBA when you can express benefits in terms of money. The reason for this is that for an expression in money I am using numbers, which is more accurate and more precise. But when you are not able to do it with a particular accuracy and precision, use the CEA. Whatever the ALARP is set up to (CBA or CEA), the loss of human lives must not happen and this is the main principle in safety!

1.8 Using the BM in Risk Management

Mostly in the petroleum and oil industry literature (dedicated to the BM) you will find the use of the BM as a connection with the controlling of the risk and management system. There is nothing strange in that having in mind that the BM deals with safety issues in petroleum and oil industry. There, the BM is used to explain all measures (named as barriers!) to improve the quality and safety of all operations. So in petroleum and oil industry, the BM is used as help to prove that all risks are not only analyzed but also managed.

Maybe some of the practitioners will find themselves confused by the BM in this book. I am using it to explain the assessment side of the BM, because when you finish with analysis, you try to find solutions (elimination and mitigation) for all possible problems (failures or factors) determined by your BM analysis. And when you implement these solutions, you are going to put them into your already produced BM and recalculate BM again to prove yourself that everything is OK. So, after the first building of the BM trees, later you change it by including solutions for elimination and mitigation just to be sure that everything will go good. I would just emphasize here that the first assessment of the system by the BM is most critical for the future functioning of the system.

Starting the BM in Risk Management can be explained in few steps. The first step is the definition of the Main Event. It is a particular event that may do harm to or damage humans and assets. This is a simple event, but it is complex by the preconditions (other events) and by their combinations that are causes for the Main Event. If I use "car crash" as a Main Event, I can notice that there are many causes for this: speed, road conditions, traffic, car maintenance, driver condition, etc. Also, these causes can interfere between them and they can create different outcomes for the Main Event. After defining the Main Event, I continue with the BM by moving left, towards the Pre-Event area that is FTA.

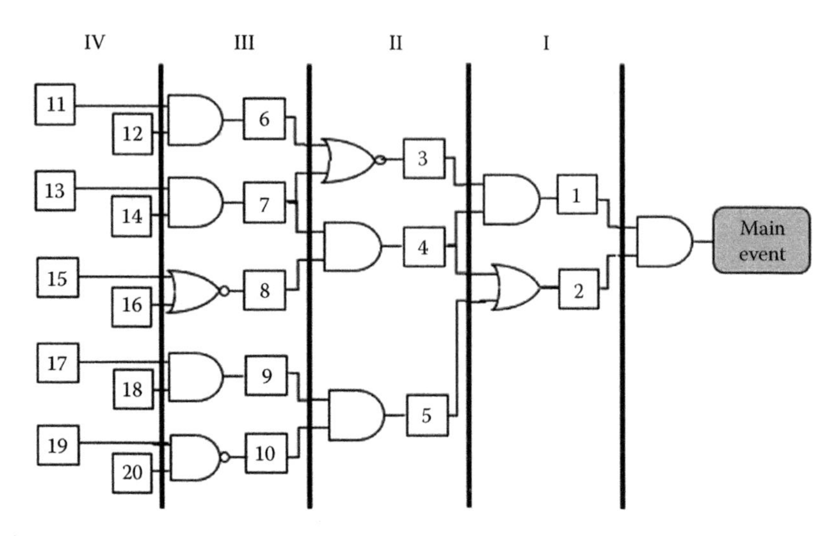

FIGURE 1.6
Iterations and events that are leading to the Main Event in FTA.

The determination of the Main Event is based on its consequences: If it can produce death, injuries, harm or damage, it must be taken into consideration. When this particular Main Event is determined, then all consequences of the Main Event happening should be determined. I may determine the Main Event only knowing one consequence, but further consideration can show other consequences. This is a team work and shall be done with a team comprising competent employees from different departments. If it is done by the safety department, then this department must have employees with different skills covering all activities of the company. The reason for this is that the particular Main Event will not have the same influence on all parts of the company. The determination of the overall influence over different parts in the company (or outside the company) is of utmost importance for supporting the holistic approach of the BM.

Starting to draw a diagram for the FTA, I deal with the second step for the risk analysis. In Figure 1.6 is shown hypothetical FTA with 20 events and 4 iterations directing to the Main Event. The diagram is drawn from the Main Event towards the nearest hazards that can produce the Main Event. It is an iterative process that starts with events that can be both individual and/or a combined cause for the Main Event. Let us call them primary* (I) events.

* In some research works you will find the name "primary" for events that are at the bottom of the FTA (most left in Figure 1.6). Actually, they define them as starting events for the further combination that lead to the Main Event. Having in mind that FTA starts from the Main Event to lower events and my definition is different due to the fact that these are the first events that I establish. So these are directly events that cause the Main Event and I name them primary events.

It is understood that these primary events can be caused by other individual events or combinations of particular events. So I draw these (so-called) secondary (II) events. And going on and on (III and IV), I can finish the drawing where there are no more causes to consider. The iterations are going from the Main Event to primary (I) events, later to secondary (II), and so on (III and IV), until I stop. The number of iterations and number of events in every iteration are determined by the nature and definition of our Main Event and structure of the system that is under consideration.

Looking at Figure 1.6, I can notice that the combination of events from 11 to 20 are contributing to the happening of events from 6 to 10; combination of events from 6 to 10 are contributing to the happening of events from 3 to 5; combination of events 3 to 5 are contributing to the happening of events 1 and 2; and finally, the combination of events 1 and 2 are contributing to the happening of the Main Event. It is important to emphasize here that the diagram in Figure 1.6 is constructed going from the Main Event (right side of the diagram) to the left side, but the calculation of the probability for the Main Event is going from the left to the right side of the diagram.

When the drawing of FTA is finished, I dedicate the particular probabilities to every event inside, and using Boolean formulas, I calculate the probability of the Main Event. When I finish this, I can look for possibilities of implementing some measures that can eliminate or decrease some of the individual probabilities. This is an activity that is supposed to prevent or at least postpone the Main Event. This is a very important step, because it will improve the overall performance of the BM. As early as I can handle the problems, the more benefit it will bring to the area of costs for solving the consequences later.

Figure 1.7 shows the so-called economic lever for solving problems during design (red lever) and during production (blue lever). The movements (presented by red and blue arrows) on the left side of the levers are costs and the movements on the right sides (presented by green arrows) are showing the effects of these costs. You can notice that solving the problems during design requires lower cost (red arrow) with a bigger effect (green arrow on another side of the design lever) than solving them during the production process (blue arrow).

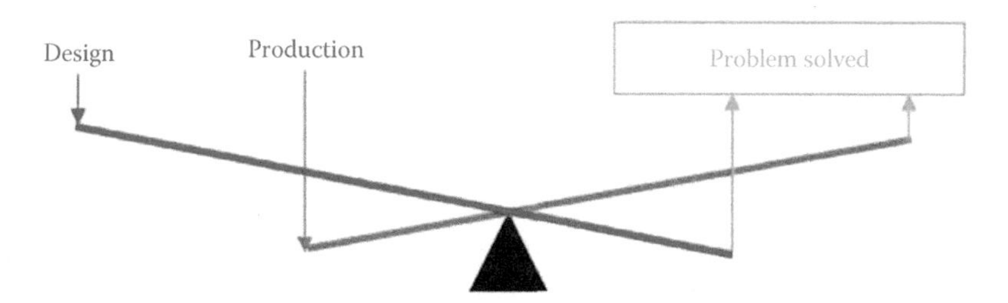

FIGURE 1.7
Economic lever for failures and costs.

To prove the diagram in Figure 1.7, I present you one simple example.

If there is a mistake in the design of the product and this mistake is found during the design process, then the rework costs in production will not be increased (because there is no production) and there are no consequences concerning the customer. But if the mistake in the design of the product is discovered when the product is sold to the customer, then the rework costs that will deal with the mistake may increase up to 10,000 times. That is the reason why implementing the BM during the design process is strongly recommended.

If the problem has safety consequences, then, the cost of fixing, hospital or fines can increase dramatically. A general rule is: The sooner I find and solve the problem, the better, because I will spend significantly less to fix it!

Speaking about costs, I can speak about one more benefit of the BM: Having in mind that FTA and ETA are connected, trying to solve some potential problems during the FTA will decrease the cost for barriers and dealing with consequences later (during the ETA).

When the FTA is finished, then the third steps starts: ETA. The main point here is that the ETA will also trigger some changes, maybe even in the FTA diagram. So, if you are thinking that the FTA and ETA are independent in the BM, you should forget it. Some of the measures assumed to be used in the ETA may be better and easier to implement if there are changes in the FTA. A simple example is the road traffic. Putting a speed limit on the road influences the FTA (the time to react will be bigger and it can prevent the crash), but also the injuries and car damages will be smaller (due to the lower speed!) and easier to handle with the ETA!

1.9 Determining the Severity of the Main Event

Severity is the level of damage that happens to humans, environment, and assets. The determination of severity should be based on an in-depth case study for a particular area of interest, extensive field monitoring, and in-depth previous event data analysis. You can use (so-called) macroregion gathered data (for the particular events that happened in all industries), but the data dealing with the (so-called) microregion (events that happened in your company) are more important. Even though there are more data in macroregion (all industries!), you execute the BM for your company (microregion!) and these data are a bigger point of interest for you!

It is important to understand that the criteria for severity are highly subjective. *Subjectivity, when we are speaking for money is coming from fact that* the risk of losing 1 million USD for a small company is catastrophic, but it is negligible for a huge company. The level of severity also depends on the nature of the company, because the consequences of the same event are not the same for a company that produces goods and for a company that offers services.

Also the severity is strongly dependent on the place where the risk takes place. The severity of bad weather in equatorial regions is different than that in the continental areas. Storms there are expressed by hurricanes, typhoons, and so on, but in continental areas, the only possible point is flooding and it happens very rarely. And the last example: A fault in the engine of an aircraft on ground will produce less severity than a fault midst flight.

Another point is that different categories of assets have different levels of damage and this damage can happen in different ways. There are also primary damages that happen instantaneously with a Main Event and there are secondary (hidden) damages that indirectly and later affect the processes and operations inside the assets. Let us say that an earthquake may not damage the equipment inside a building, but it may damage the building in a way that it is not safe anymore for the usage of production equipment. It means that the severity of consequences, measured in time and money, will be bigger than money needed to build a new building (due to time and money to transfer production equipment into a new building).

The social situation and the level of education of employees are also very important factors in classifying the severity even for the same industries. That is the reason why every industry usually has some criteria for severity of consequences, and depending on the industry, they can be obligatory or nonobligatory. Anyway, every company may (and should) establish its own criteria that must not be lower than the particular regulation (if it exists) for this specific industry.

As I mentioned before, different industries have different criteria about the severity in Risk Assessment. Table 1.2 (Section 1.6) shows and explains the criteria for the severity of events in aviation. There are five severity criteria: Catastrophic, Hazardous, Major, Minor, and Negligible. For each of these criteria, there is an explanation that gives us some clarification on how to classify the event.

In nuclear industry there is a classification of the severity of unwanted radiation that also has five levels that have similar names: Catastrophic, Major, Moderate, Minor, and Negligible. Anyway, the definitions of the severity of them (in aviation and nuclear) differ due to the nature of the operations there. The criteria for aviation are given in Table 1.2 and the criteria for radiation severity are given in Table 1.3.

The categorizations of severity with five levels of consequences are most common in all industries. Nevertheless, there are some industries with only four levels. Let us say MIL STD 882E, which is USA Ministry of Defense standard for System Safety, establishes only four criteria: Catastrophic, Critical, Marginal, and Negligible.

The general rule of the thumb is that whatever the established criteria for severity are, the most serious consequence should be prioritized during the calculation and mitigation of the risks! It means that with car collisions

TABLE 1.3

Consequences of Nuclear Radiation in mGy

Level of Consequence		Annual Dose (mGy)	Reasons
5	**Catastrophic**	>50,000	Skin necrosis
4	**Major**	2,000–50,000	Below erythema level
3	**Moderate**	150–2,000	500: legal limit; <2 Gy early transient erythema—ICRP 85
2	**Minor**	50–150	<150: nonclassified level
1	**Negligible**	<50	(100 mGy) Not clinically relevant functional impairment—ICRP 103/Also including non-radiation worker (50 mGy)

Note: mGy is an abbreviation for miliGrey (1000 times less than Grey, unit for ionizing radiation in International System of Units [SI]).

I should prioritize the calculation of the risk with the severity of human casualties and later deal with the severity of material damage of the cars or assets around the crash site.

1.10 Benefits of BM

The benefits of the BM are huge. As I mentioned in Section 1.5 (Risk Management), this is a methodology that provides a holistic approach to Risk Management by identifying the hazards, quantifying them into risks, and helping to build barriers for elimination and mitigation of the consequences.

Starting from the beginning, with the BM, I can consider all hazards that can contribute to the happening of a particular event and determine the combinations between them. During the Pre-Event Analysis (FTA) I can describe different scenarios for the Main Event to happen and evaluate them by quantifying the probabilities of something like that to happen. Looking at the FTA diagram I can better understand the influences of the failures in the system and determine which of scenarios is most critical. It will later help me to establish particular elimination and/or mitigation measures to decrease the frequency and consequences of particular risks.

The same point is also applicable for the Post-Event Analysis (ETA) where I can produce all of the credible scenarios of the development of the consequences that will help me to build barriers to eliminate or mitigate them. Using the ETA I can even calculate the probabilities for some particular outcomes (Failure or Success).

For trained people this methodology is an excellent tool for qualitative and quantitative Risk Assessment and it helps to produce effective and efficient

Risk Mitigation. It is flexible in a way that it allows safety managers to choose the scope and details of the analysis based on real situations in their system. When implemented during the design phase, it can bring a good picture for the product about both safety or quality issues and it can contribute to preventing and decreasing possible economic losses that can arise during the warrantee period.

But there is another side of the BM.

BM is an excellent tool for industry, but not for scientific purposes. FTA for scientific purposes has a lot of deficiencies and is often changed with the Bayesian Network approach, which is advanced, particularly for dynamic environments. Unfortunately, the level of math knowledge necessary to execute it is not convenient for production processes.

Another point is that this is a model approach. I am building a model of my system under consideration that represents my perception of my system. But I will speak about that later.

2

Probability*

2.1 Introduction

To use the BM you should understand probability. The BM is known as a part of the so-called Probabilistic Risk Assessment approach due to the use of probabilities for quantitative calculations.

Probability is part of mathematics that deals with the uncertainty of random events. It is founded in seventeenth century in France and we may consider French mathematicians Blaise Pascal and Pierre Fermat as its "fathers." During that time it started as a mathematical background on gambling, but in eighteenth century it moved forward as science.

Probability deals with random events,† but there is nothing random in nature. By our understanding, the events that we cannot fully understand, determine, or calculate are called random events (or events by chance!). So, when I choose something without any kind of criteria,‡ I make a random choice.

Probability is part of our everyday life, even though we are not always aware of it. In planning our everyday activities, we use knowledge gathered through education and experience. My experience taught me that things do not always happen the way I had planned them, and the real reason for that is as follows: I have a lack of knowledge that can explain how those things are functioning, so my assumptions do not always fit reality. Whatever the level of our education and experience is, there are things in nature that are so complex, and very often, we do not know the true mechanism of how these things happen. So, in the absence of enough knowledge and understanding, we try to predict the outcomes of our activities, especially for the things that we know that our knowledge (or information) is not enough. It means that

* Please note that I will stick to probability that is strongly connected with the application for safety cases.

† As the antipode of ***random events*** there are ***deterministic events***. Normal processes in industry are deterministic events (intentionally made to function that way!).

‡ I choose it, but I choose it without any reason, just by chance!

we use probability after we statistically investigate the previous events, but it will not help us to accurately predict them in the future.

Tossing a coin is connected with the laws of physics, and scientifically speaking, to calculate the outcome ("head" or "tail") I need to take into account a lot of laws. These laws are applicable to the conditions that are viable at the moment I toss a coin. These conditions are not known to us, so we need to measure and implement them into the laws of physics in real life. Some of the conditions that are important are as follows: the law of gravity at the place of tossing the coin, the force used to toss it, the angle of tossing, the mass of the coin, the distribution of mass in the coin, the angle of the coin hitting the ground, and so on. In theory if I have all these data I can calculate the outcome, but in reality I do not have the equipment (sensors) to measure all of these data and powerful computers that will calculate all of these calculations in real time. So, I say that the outcome of the tossing of a coin is a random event and I can only guess it.

Today, speaking about safety I can notice that it is similar to gambling. Actually, almost every day that I drive in my car is "gambling" because there are incidents and accidents on the streets that can always happen. Probability plays a very big role in safety. Having in mind that I am actually predicting future events, I am putting myself in a situation to make decisions based on previous events and situations. I have registered these events and situations by my own experience or I gather information regarding them from different sources (radio, newspapers, TV, word of mouth, etc.). That is the reason why these decision makings (predictions) must not be just provisional, meaning that I have already gathered some knowledge by information that I have registered. Anyway, these predictions (and their background information gathering) should be based on science. And as a science for these decision makings (predictions), I am using probability. Probability, as a part of mathematics, is a tool to be more effective and efficient in those activities.

As I said before, I cannot always use probability. The reason for that is the necessity of reliable data that I can only gather if there are considerable numbers of previous events. Having in mind that I am striving for my life to be without any harm and damage and bad things do not happen very often, reliable data might not be easy to gather. Another problem is the variability of situations. Tossing a coin can be treated as pretty much an invariable situation, but accidents can happen in plenty of different ways, even though the consequences can be the same. So, because I might not have enough data for safety predictions, I use frequency instead of probability.

The newest development of safety is establishing different views of safety. Instead of dealing with safety as an activity looking and preventing "what is going wrong" (Safety-I), there is a proposal that safety should actually be investigating the "failure to going right." This is so called Safety-II and it is based on "what is going right." Safety-II states that by improving the situations of normal (right) functioning of the systems, we actually decrease the chances of an abnormal (wrong) functioning of the systems. It is defined as an ability to provide success under different and dynamic conditions. Safety-II states that with today's complex systems it is better to strive to maintain the normal operations instead of looking for possible causes of malfunctioning. Having in mind that Safety-I is also based on the combination of different conditions, calculating the frequency (probability) of the risk should take into account different situations. All these situations in today's complex systems will create complex interdependencies between the elements of these complex systems.

2.2 "Black Swan" Events

Let me speak about events that are part of this chapter only by their specifics because when they happened, that was the first time they happened. The probability of an attack in the USA on September 9, 2001, was zero because such a thing had never happened before. The probability of having suicide bombers who blow themselves just with the intention to kill other people was zero in the last century. These events never happened before and nobody assumed that they might happen today.

The probability of the Fukushima Daiichi disaster to happen on March 11, 2011, was also zero due to the same reason. The probability of the Vajont Dam disaster, October 9, 1963, in Italy was also zero. Mathematically speaking, I cannot determine the probability of such events because they never happened before, so no calculation is possible there. Nevertheless, it was assumed that Fukushima Daiichi and Vajont Dam accidents might happen and particular modeling simulation produced some preventive measures, but obviously the simulations were not enough.

These types of events are known as ***Black Swan*** events. These are events that happened as a total surprise and no one could have predicted such a situation. There are no data for them and they look almost impossible to be predicted. This is one good example where the probability of an event to happen was zero, but the event happened anyway. These events are so strong that the consequences are also terrible. Due to the surprising nature of Black Swan events, the risk assessment is possible to be done only with a poor integrity modeling.

One solution is to implement resilience engineering or, in other words, to make our systems resilient for such events. This is a solution, but unknown variables are too many and assumptions are too weak. Producing systems resilient to such an event can be very expensive...

I have read somewhere that today, actually, the probability that your system will be destroyed by such an event is even higher than the probability of it to be destroyed by known events. The reason for this is that a good safety management system takes care for known events, but it does not prevent Black Swan events.

2.3 Sets

For probability I use sets, because the elements for calculations abide to the laws and rules of the mathematical theory of sets.

There is one important definition in the mathematical theory of sets and it is the ***universal set*** (β*). Using simple words, it is the aggregation of all the elements (members or participants) in anything under consideration. Let us say that all letters in English alphabet are the universal set of letters used to write and speak English language.

In addition there is also the definition of an ***empty set*** (Ø). It is a set with zero elements inside (empty!).

If I have one particular set with few elements inside, I can determine some so-called subsets inside this set. Let us say that the set of all outcomes of tossing a dice is {1, 2, 3, 4, 5, 6}; then the subset of odd outcomes will be {1, 3, 5}, and the subset of even outcomes will be {2, 4, 6}. In addition, I can look for the outcomes of tossing a dice that is odd and bigger than 2 and this subset will be {3, 5}. I can produce an infinite number of different rules that will produce infinite subsets inside the set of outcomes when tossing a dice. Even though the rules can be different, it is no strange subsets to be the same. Typical examples for these are the rules "All prime numbers" and "All odd numbers" from tossing a dice outcomes. The rules are completely different mathematically, but the subsets will be the same: {1, 3, 5}.

I can also produce new sets by adding different sets. The addition (summation) of different sets will produce another set that is called ***union*** (∪). To further clarify, sets that are used in the sum become subsets of a new set called union.

I can also produce an ***intersection*** (∩) between two sets. An intersection of two sets is made up of elements that are common for both sets.

Figure 2.1 presents the so-called Venn diagram that is visualization of an empirical presentation of union and intersection between sets A and B. As I can see in the Venn diagram, there are two sets $A = \{1, 3, 5, 6, 7, 8\}$ and $B = \{2, 4, 6, 8\}$. Looking at the diagram, there is a union $A \cup B = \{1, 2, 3, 4, 5, 6, 7, 8\}$, presented by the violet line and intersection $A \cap B = \{6, 8\}$ presented by elements that belong to both sets (red arrow points to the area with these elements).

There is another term that needs to be defined here and this is ***complement.***

As shown in Figure 2.1, there is a subset that is given as an intersection $A \cap B$ of the union $A \cup B$ and for this intersection $(A \cap B)$ complement $((A \cap B)^C)$ are all elements that belong to A and B, but do not belong to $A \cap B$. In other

* In different books you can find different symbols for the universal set (U, Ω, etc.).

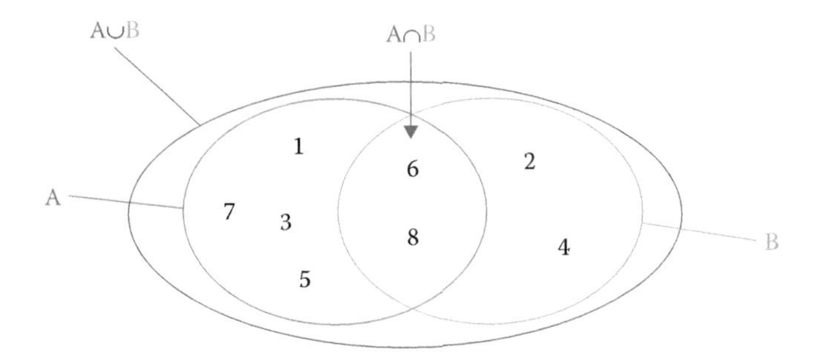

FIGURE 2.1
Venn diagram for the union and intersection of two sets.

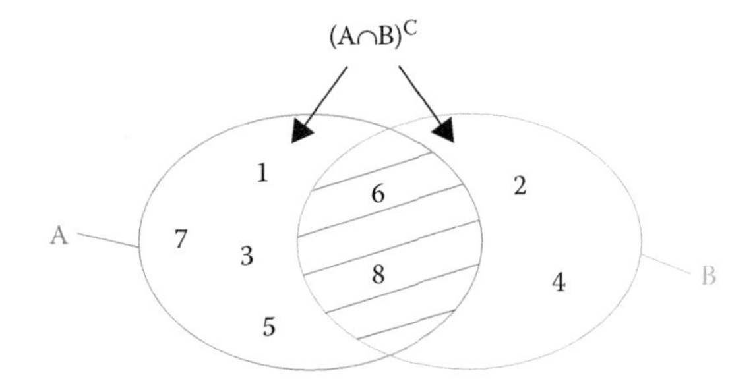

FIGURE 2.2
$A \cap B$ is marked with red lines and $(A \cap B)^C$ pointed by black arrows.

words, $(A \cap B)^C = \{1, 3, 5, 7, 2, 4\}$ (Figure 2.2). In Boolean algebra and engineering, complement is used as negation (not A or not B) and in the literature you can find the following symbols for A^C, A' or $\bar{A}$.

In our earlier example (Figure 2.1) if I speak about $A \cup B$, I can write:

$$B + B^C = A + A^C = A \cup B$$

Of course, if $A \cap B = 0$, then $B^C = A$ and $A^C = B$.

Speaking about complement, I can connect it with another operation of sets: ***subtraction***. It is written like this:

$$A - B = \{1, 3, 5, 7\}$$

Or in other words, set A – B is made up of all elements that belong to A and do not belong to B. Looking at Figure 2.2 you can notice that there are elements like the ones that are given by the previous formula.

TABLE 2.1

Algebra Laws for Sets

Commutative laws	$A \cup B = B \cup A \quad A \cap B = B \cap A$	
Associative laws	$(A \cup B) \cup C = A \cup (B \cup C)$ $(A \cap B) \cap C = A \cap (B \cap C)$	
Distributive laws	$(A \cup B) \cap C = (A \cap C) \cup (B \cap C)$ $(A \cap B) \cup C = (A \cup C) \cap (B \cup C)$	
De Morgan's laws	$\left(\bigcup_{i=1}^{n} A_i\right)^C = \bigcap_{i=1}^{n} A_i^C$	Complement of the union of all sets is equal to the intersections of all individual complements of the sets.
	$\left(\bigcap_{i=1}^{n} A_i\right)^C = \bigcup_{i=1}^{n} A_i^C$	Complement of the intersections of all sets is equal to the union of all individual complements of the sets.

All of the well-known mathematical laws for calculations are also applicable to sets' operations (Table 2.1), which allows us to do different mathematical operations with sets.

2.4 Sets and Probabilities

In mathematics, different disciplines have different universal sets. Let us say, for calculus, that the universal set consists of all real numbers, for complex analysis we have all complex numbers, and so on. For our case (dealing with probability connected with risks) our universal set will be a set of all events that may trigger the Main Event. Of course, this set will be established by us and it can be different for each Main Event.

Whatever the universal set is, each element in this set will have a particular probability of happening. I can even establish a set of probabilities of the elements in the universal set of events causing the Main Event. Having that in mind, the values of all probabilities must be equal to 1 (universal set, $\beta = 1$). I can use 1 instead of β when speaking about calculating all the probabilities of the Main Event to happen. What is important here is to understand that sum of calculations of all probabilities of known primary events* should always result in 1.

* Primary events are causes of Main Event on Figure 1.6, Chapter 1.

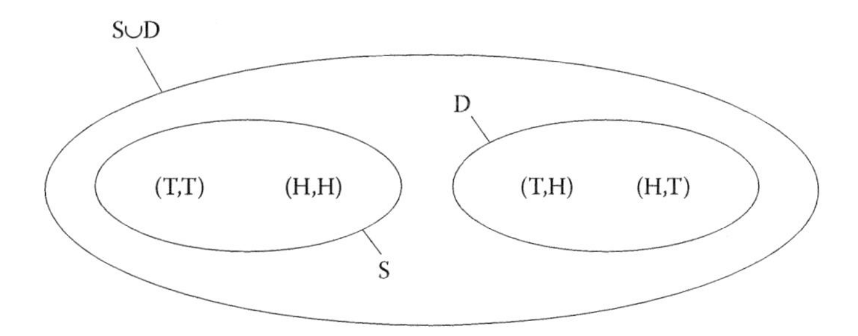

FIGURE 2.3
Venn diagram for the union of two subsets S and D for tossing two coins.

Combining (for the purposes of this book) sets and probabilities, I can say that for my probability experiments, the ***universal set*** will actually be ***Sample Space*** (SS). SS in probability is the set of all elements (in this case called outcomes) for a particular experiment. The number of elements in SS for the tossing of a coin is 2: tail (T) and head (H). Similarly, the number of elements of the SS made by the outcomes from throwing a dice is 6 (1, 2, 3, 4, 5, and 6).

What is important to mention here is that all the different SSs* follow the same rules regarding the sets and combinations of sets (union, intersection, complement, subtraction). It means that I can add SSs and I can make different sets or subsets using specific elements (events) of different SSs.

Speaking about the empty set (Ø) in the context of probability, I can say that it is a set with 0 elements inside (empty!), which means that the probability of this to happen is equal to 0. If I define the value of a universal set in probability to be 1, then the probability of this to happen is 1 (it will happen always!). The connection between the empty set (Ø) and the universal set (1) in probability is giving us the so-called first axiom in probability: All of the values of the probabilities in one set must belong to the values between 0 (Ø—empty set) and 1 (β—universal set) and their sum must also be 1.

Looking at Figure 2.3, from the SS of the results of tossing two coins, I can create two different SSs (or subsets). One subset (S) will be made up from elements that give us the same results and the other subset (D) will be made up from elements that give us different results (different outcomes) when tossing a coin—T and H, H and T. So I will have for S and D:

$$S = \{(T,T),(H,H)\} \qquad D = \{(T,H),(H,T)\}$$

The union of S and D is presented by the following formula, which is an empirical presentation, but in Figure 2.3, I present a Venn diagram that is a visualization of the empirical presentation.

$$S \cup D = \{(T,T),(H,H),(T,H),(H,T)\}$$

* SSs is plural for SS (**S**ample **S**paces = SSs).

The union and intersection created by these two sets (S and D) have great meaning in the case of probability.

They are connected with a so-called second axiom of probability that deals with ***mutually exclusive*** (independent) and ***not mutually exclusive*** (dependent) events.* Mutually exclusive events are events that do not rely on each other or there is no connection between them. Also they usually do not happen at the same time. For example, if I toss a coin, I cannot get "head" and "tail" at the same time. The simplest situation of mutually exclusive events is presented with the tossing of a coin: Whatever the result of the first toss is (H or T), the second toss will produce independently again H or T. Speaking about probability, any of the events may happen independently and there is no connection between them.

The intersection in probability does not present mutually exclusive (dependent) events. In Figure 2.1, elements 6 and 8 belong to set A and set B at the same time, which means that for situations with these two elements, events A and B are dependent: They can cause A or B to happen at the same time. When speaking about the probability of dependent events, both events must either happen in a row (first one and then another one) or both at the same time (happening of the first event will trigger happening of the second one).

The concept of complement of sets is also important for the probability in the BM, because there are two subsets of operations in any system. The first one is a subset of all possible ways that provide normal operations of the system and the second one is a subset of all the faults that can happen, so normal operations are not possible anymore (the system is abnormal). Speaking about probability, I can say that there can be a probability for normal operation given as P(n) and there is also a probability of having a faulty system† (abnormal operation) given as P(f). Having that in mind, in this context, there is no other possibility (system is either normal or faulty) I can write:

$$P(n) + P(f) = 1$$

In other words, probabilities for normal and faulty (abnormal) operations relate to other as complementary sets of probabilities for normal and faulty operation:

$$P(n)^C = P(f) \text{ and } P(f)^C = P(n)$$

We can infer that knowing the probability for normal operations can help us find the probability of a faulty operation. And vice versa, knowing the

* Do not mix dependence with correlation. Correlation is a part of dependence and it means that both events change dependently (if one increases, other increases (or decreases) at the same time).

† A faulty system is a system that does not work properly or it does not work at all. In other words, it executes an abnormal operation.

probability of having a faulty operation we can find the probability for a normal operation of the system. It is explained by:

$$P(n) = 1 - P(f) \text{ or } P(f) = 1 - P(n)$$

The newest development on Safety-II is using this situation with probabilities. Instead of striving to decrease the number of faulty system situations, we can try to improve (extend) the normal operation of the systems. And vice versa, by putting efforts to sustain a normal operation of the system, I am actually decreasing the number of situations when the system might turn faulty. I will say more on this later.

2.5 Basics of Probability

To use probability, it is very important to know the rules to calculate the total probability of the Main Event if there is a combination of primary events with individual probabilities. Finding the individual probability of an event should be strongly scientific and it can be done through an experiment. Tossing a coin is one experiment that can be repeated many times, and it helps me to establish the values of the probabilities for "heads" and "tails." During the experiment, I register all possible outcomes by repeating the experiment and count how many times each of them happened.

In general, having more elements (more outcomes) in a particular experiment (SS) poses a requirement for a higher number of tries of this experiment, so you can get reliable results. It means that the number of tries for tossing a coin (two outcomes) is considerably smaller than the number of tries for tossing a dice (six outcomes).

The number of tries is connected by the definition of frequency and probability that are given by the following formulas:

$$F(A) = \frac{n}{N} \qquad P(A) = \lim_{N \to \infty} \frac{n}{N}$$

where F is the frequency and P is the probability, A is the event that I expect to happen, n is the number of times event A has happened, and N is the number of total tries during which event A could happen.

You can notice that they differ only by the number of tries. If this number is finite and not so big, I consider the frequency of the event happening. But if the number is very big and it approaches infinity, then I speak about probability. Having that in mind, I can simply say that the frequency of an event happening (from a plenty of outcomes) is actually a rough estimation of the probability of this event happening.

A simple explanation of the formulas from the previous is as follows: If I toss coin N times, I will get a "tail" n times (A). Then, the frequency of getting a "tail" is n/N. If N is big (approaching infinity), then the frequency becomes the probability.

In real life I do not need to go as high (to reach infinity) to have probability. It is enough to deal with the range of 100,000 tries, of course, if it provides the consistency of the conditions that are connected with the event. The number of 100,000 tries is just an example. The number of tries should be big enough, so the further tossing of a coin will not change the result significantly. Significantly in this situation means that changes of the results are possible only on the third or fourth or tenth decimal place, depending on the case. In our case, when tossing a coin, you can notice that 1,000 tosses will not change our result that will be very close to 0.5 (for both "tails" and "heads").

The earlier explanation is about probability that can be calculated intentionally. I just take a coin and toss it 100,000 times. The results will help me calculate the values for "heads" and "tails" frequencies. But in our (safety) case it is not always applicable. In safety I am concerned with "something which does not need to happen" and this is the fault of an equipment or failure of an operation. So in safety case, instead of looking for data regarding "success," I am looking for data regarding "failures." And I cannot and may not try 100,000 times to check the probability of a passenger surviving a car crash. Simply, it is not allowed. Data available for these cases should be found from regulatory bodies that deal with statistics. I will not put a clear boundary between frequency and probability in this book. I will mostly speak about probability just to emphasize the scientific background of the BM, but in your analysis you will have to use the word "frequency" very often due to the lack of appropriate amount of data.

Let us continue.

The SS for tossing a one coin* can be written as:

$$SS(1) = \{T,H\}$$

And the SS for tossing two coins will be:

$$SS(1+1) = \{(T,T),(T,H),(H,T),(H,H)\}$$

But this SS will exist if I toss the coins one by one (first one and then another one). For this case, this is the same as if I am tossing one coin twice. If I toss two coins in at the same time, SS will be:

$$SS(2) = \{(T,T),(T,H),(H,H)\}$$

As you can see, there are only 3 elements, because I cannot know which coin is H and which coin is T. So, (H, T) and (T, H) will be the same.

* H is for "heads" and T is for "tails."

The probabilities of the outcomes from tossing one coin will be:

$$P(H) = 0.5 \qquad P(T) = 0.5$$

In words, the probability of getting H (head) will be 0.5 (50%) and the probability of getting T (tail) will be 0.5 (50%) also.

For tossing two coins one by one, the probability for every outcome will be:

$$P(T,T) = 0.25 \quad P(T,H) = 0.25 \quad P(H,T) = 0.25 \quad P(H,H) = 0.25$$

The reason for these values is the probability of H or T for one coin that is 0.5. Having in mind that the tossing of two coins (one by one) is totally independent (results from the first tossing do not affect the result of the second tossing) the probabilities will be multiplied so:

$$P(T) \bullet P(T) = P(H) \bullet P(H) = P(T) \bullet P(H) = P(H) \bullet P(T) = 0.5 \bullet 0.5 = 0.25$$

The probabilities of two coins tossed together will be:

$$P(T,T) = 0.25 \qquad P(T,H) + P(H,T) = 0.5 \qquad P(H,H) = 0.25$$

As you can notice, P(T, H) is 0.5 because there is no difference between (T, H) and (H, T): Every toss with this result will have two more chances to happen and can be presented with a sum.

From the earlier values of the probabilities I can notice that the overall probability of all elements (outcomes) in a particular SS (experiment) must be 1. It is always valid only if the outcomes of the experiment are independent (mutually exclusive events), which is true in this particular experiment (tossing a coin). If the overall probability is not 1, then there is a mistake. Some of the reasons for the mistake could be:

- The number of tries is small (so I am missing some of the outcomes).
- The events are dependent (so I am making mistakes when calculating the probability).
- I have wrong assumptions about the outcomes.

Let us explain this in detail.

In the past (before the 9/11 attacks), the probability for such an attack was 0. Starting with terrorism, there were no data that terrorists would be willing to kill themselves to execute an act of terrorism. So, having in mind that such a situation has had never arisen, no one assumed that it is even possible (Black Swan event). There were some statistics about the terrorist attacks and it shows the probability of every type of attack. The overall sum of the probabilities was 1. But after 9/11 things changed: New "outcomes" in the "sky of terrorism" arose and because of that a new probability for this new outcome was added to the calculations. It means that this probability affected all other probabilities in every other type of terrorist attack, so again, all probabilities needed to be recalculated. Today, using statistics I can see that suicidal terrorist attacks are most common (only in 2015 and 2016, we had terrorist attacks in Paris, Brussels, Ankara, Istanbul, Bagdad, etc.).

Similar situations arose after the Fukushima disaster (2011, Japan). No one calculated the hazards of such a big tsunami to a nuclear plant. Obviously, such a big tsunami never happened before and people thought it could not happen (Black Swan). To be honest, there were some calculations, but they produced a margin of only 5.7 m (nevertheless, the method for this calculation is not known). But the tsunami had a height of 14–15 m that day. After the Fukushima disaster, additional calculations (and recalculations) of risks were performed and new mitigation measures were added to all nuclear plants in this area.

Knowing that, I emphasize, the particular value of the probability for every element in the set must be between 0 and 1.*

If the probability of a particular event is 1, it means that it will happen during every try, and if probability is 0, it means that it will never happen. But there is something interesting here: In reality, even the events with probability equal to 1 may not happen, and even events with probability equal to 0 may happen (Black Swan event). This is connected with some extremely rare events that can be registered only if I increase the number of tries in the experiment beyond the reasonable value.†

So, probability is a strongly mathematical prediction that is calculated using statistical theory for the outcomes of experiments. If I do not register some outcomes during the experiment, then I will write it as having probability equal to 0, which does not mean that it cannot happen. Or, if I toss a coin 10 times and every time I get "heads" (it can happen!), I should not assume that the probability for "heads" is 1, because that is wrong. Wrong simply because 10 tosses are not enough for a credible result. In September 2009, the Bulgarian lottery selected the winning numbers 4, 15, 23, 24, 35, 42. Even though the probability for it to happen again the next week was 0, the same numbers were selected again next week.

And these examples are hiding some very interesting points about safety: Even though I can use the BM to analyze the safety and to help to eliminate or mitigate the known risks, the main challenge is to deal with unexpected incidents or accidents (things that have probability to happen equal to 0).

* Very often probabilities are shown as percentage, so 0 probability will be 0% and 1 (full) probability will be 100%.

† Black Swan events are explained in Section 2.2.

2.6 Dependent Events

Not always will the outcome of an event be independent. Car accidents are more common when the weather is rainy or snowy. This means that there is an interconnection and dependency in the probability of car accidents and weather conditions. These are cases when I can speak about conditional probability. This is a probability that changes the results of the outcome of an event, if there is another event dependent on the event of interest. Simple example: If I am eating in the restaurant and I am not satisfied with the service there, my next visit to the same restaurant is highly improbable. My previous visit to the restaurant affects the probability of my next visit to happen. This depicts conditional probability where events are dependent. When having two dependent events (A and B), I can write their probability as:

$$P(A \cap B) = P(A) \bullet (B|A) = P(B) \bullet PA|B)$$

where:

A and B are dependent events
$P(A \cap B)$ is the probability of A and B happening one after another (dependent events)
$P(A|B)$ is the conditional probability for A to happen if B already happened
$P(B|A)$ is the conditional probability for B to happen if A already happened
$P(B)$ is the probability of B to happen
$P(A)$ is the probability of A to happen

Transforming these formulas I can get:

$$P(A|B) = \frac{P(A \cap B)}{P(B)} \text{ and } P(B|A) = \frac{P(A \cap B)}{P(A)}$$

The first formula can be used to calculate the conditional probability of A to happen if B has already happened. The second formula can be used to calculate the conditional probability of B to happen if A has already happened. Do not forget that A and B are dependent, so $P(A \cap B)$ is the probability of both (A and B) happening in a row or at the same time.

Anyway, if two events happen one after another or they happen at the same time, it does not mean that they are dependent! So, be careful when determining which events are dependent and which are independent.

Mathematically speaking,* I can say that A and B are mutually exclusive (independent) only if:

* Using the language of probability, dependent events are "not mutually exclusive" and independent events are "mutually exclusive." In the book I will mostly use the terms "dependent" and "independent."

$$P(A|B) = P(A) \text{ and } P(B|A) = P(A)$$

whatever P(B) and P(A) are!

Let us mention the formula for the probability of any of two dependent events (A and B) to happen by using sets:

$$P(A,B) = P(A) + P(B) - P(A \cap B)$$

For three independent (mutually exclusive) events (A, B, and C) the overall probability of **any one of them** happening is:

$$P(A,B,C) = P(A) + P(B) + P(C)$$

But if they are dependent, the probability for any one of them to happen will be:

$$P(A,B,C) = P(A) + P(B) + P(C) - P(A \cap B) - P(A \cap C) - P(B \cap C) + P(A \cap B \cap C)$$

Looking at the previous formula, you can notice that for independent events, the overall probability is the maximum of the probability if the events are dependent. Let us explain how I got theses minuses in the formula.

For this purpose I will use a Venn diagram for three events (A, B, and C).

I have shown in Figure 2.4 the Venn diagram of a union of three sets (three probabilities of events) A, B, and C, which are intersected. It is presenting three events which are dependent (not mutually exclusive) for some of their elements. I can write:

$$A = \{1,3,5,6,7\} \qquad B = \{2,4,6,7\} \qquad C = \{5,7,8\}$$

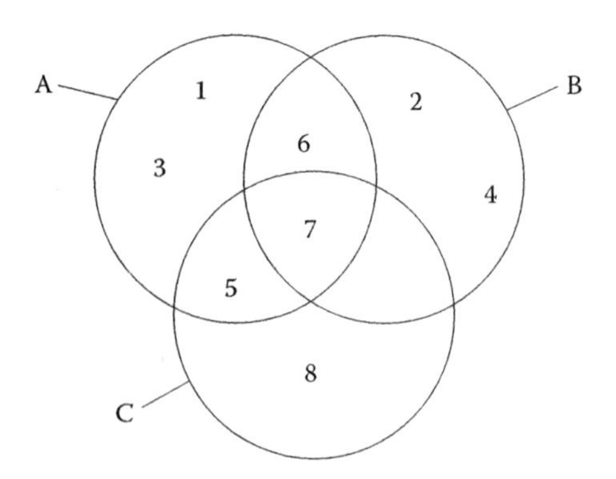

FIGURE 2.4
Venn diagram for the union of three sets with intersection (dependent).

Having in mind that there is union of them (of any event to happen!) the probability will be:

$$A \cup B \cup C = \{1,3,5,6,7\} + \{2,4,6,7\} + \{5,7,8\} = \{1,2,3,4,5,5,6,6,7,7,7,8\}$$

As you can notice that some of the elements in the union are repeated, simply because they are part of intersection. So, to have a reasonable result I must not take them for calculation or, in other words, I need to subtract them from the union. And that is the reason for the minuses in the formula for dependent events.

For few independent events (A, B, and C) the overall probability of **all of them** happening at the same time or one after another is:

$$P(A,B,C) = P(A) \bullet P(B) \bullet P(C)$$

If they are dependent, then the formula for the probability of all of them happening at the same time will be:

$$P(A,B,C) = P(A) \bullet P(B|A) \bullet P(C|A \cup B)$$

Or expressing the formula in words: The probability of all three (A, B, and C) dependent events happening at the same time (or one after another) is the product of the probability of A happening, the probability of B happening (if A already happened), and the probability of C happening (if A and B already happened).

2.7 Bayes' Theorem*

There is a theorem that helps us find the probability of dependent (not mutually exclusive) events. This theorem is presented by the formula:

$$P(A_k|B) = \frac{P(B|A_k) \bullet P(A_k)}{\sum_{i=1}^{n} P(B|A_i) \bullet P(A_i)}$$

Let us explain this formula.

* There are different names for Bayes' theorem, so you can find it in the literature as Bayes' formula or Bayes' rule.

Speaking about a car, I can say that there are plenty of subsystems making the system (car!) to function properly. There is an engine (system A_1), a subsystem for fuel storage and distribution (subsystem A_2), a steering subsystem (subsystem A_3), an electrical subsystem (subsystem A_4), and plenty of other subsystems (systems A_i). Defining the universal set for car as USC, I can write:

$$USC = A_1 \cup A_2 \cup A_3 \cup \ldots \cup A_n = \sum_{i=1}^{n} A_i$$

All of these A_i's can be presented by a union of normal situations (N_i) and failure (abnormal) situations (F_i). So, I can write:

$$P(A_i) = P(N_i) + P(F_i) \text{ and } P(A) = \sum_{i=1}^{n} P(A_i) = \sum_{i=1}^{n} \left[P(N_i) + P(F_i) \right] = 1$$

The union of all failure (abnormal) situations can be presented by B, so I can write:

$$B = F_1 \cup F_2 \cup F_3 \cup \ldots \cup F_n = \sum_{i=1}^{n} F_i$$

where:

F_1 are failures of the subsystem A_1

F_2 are failures of the subsystem A_2 and so on

In general, I can write:

$$F_i = B \cap A_i$$

Having this in mind, changing in the previous formula, I can write:

$$B = \underbrace{(B \cap A_1)}_{F_1} \cup \underbrace{(B \cap A_2)}_{F_2} \cup \underbrace{(B \cap A_3)}_{F_3} \cup \ldots \cup \underbrace{(B \cap A_n)}_{F_n} = \sum_{i=1}^{n} (B \cap A_i)$$

In words, every car subsystem (A_i) has different failures that present the failure of the car (F_i) and all of them (together) are B, which depicts situations when the car (system) cannot be driven due to a failure. Speaking about probability I can write:

$$P(B) = \sum_{i=1}^{n} P(F_i) = P\left(\sum_{i=1}^{n} (B \cap A_i) \right)$$

Going back to the formulas in Section 2.4:

$$P(A|B) = \frac{P(A \cap B)}{P(B)} \qquad P(B|A) = \frac{P(A \cap B)}{P(A)}$$

For the subsystem A_k, I can write:

$$P(A_k \cap B) = \boxed{P(A_k|B) \bullet P(B) = P(B|A_k) \bullet P(A_k)}$$

Combining the second and third item from the previous formula, I can write:

$$P(A_k|B) = \frac{P(B|A_k) \bullet P(A_k)}{P(B)} = \frac{P(B|A_k) \bullet P(A_k)}{\sum_{i=1}^{n} P(B|A_i) \bullet P(A_i)}$$

This is how Bayes' theorem was proved. You can notice that this theorem helps us find conditional probabilities speaking about dependent events. In our case, Bayes' theorem helps us calculate the conditional probability of individual car subsystem (A_k) to fail, if there is failure of car (B happened!).

Bayes' theorem in its simplest form is actually a conditional probability of two dependent events X and Y* and it is given by the formula:

$$P(X|Y) = \frac{P(Y|X) \bullet P(X)}{P(Y)}, \text{ which comes from } P(X|Y) \bullet P(Y) = P(Y|X) \bullet P(X)$$

This formula helps me calculate the conditional probability of X to happen if Y has already happened. P(X) is the probability of X happening alone, P(Y) is the probability Y happening alone, and P(Y|X) is the probability of Y happening if X has already happened. What is interesting is the fact that P(X|Y) and P(Y|X) are connected with the same formula, so the formula can be used in both directions.

Bayes' theorem is presented in this paragraph not only for the calculation of dependent events, but also because it can be used to calculate the probability

* I am using X and Y here only with intention not to confuse you by A and B that are used before explaining Bayes' theorem.

of the Main Event happening, taking into consideration the particular causes (primary events) from FTA. The sum in the formula is connected with the probability of B happening if a different A_i happened.

2.8 Distribution of Outcomes

There is another point connected with probabilities.

As I mentioned previously (Section 2.3), there is a particular SS that is actually the set of all possible outcomes for a particular experiment. But the number of outcomes is different for different universal sets. In general, more complex systems have more outcomes in their universal sets (or they represent bigger SSs). What is also important is that distributions of these multiple outcomes can be different.

If I try to hit the center of a target using bow and arrow and I register all my tries, mathematically, I can present all my progress with the formula and curve presented in Figure 2.5. This is the most common distribution and μ in the formula is the average value (my target!) and σ is the standard deviation (distance from the target when I miss the target).

There are plenty of other distributions, but most of those used in probability are variants of the Normal distribution (Weibull, Exponential, Student's T, etc.). Some of them are continuous as the Normal distribution and some of them are discrete (as Poison and Binomial distribution). Binomial distribution is one which applies to coin tossing.

In the theory of probability, calculating probabilities is strongly connected with the distribution of outcomes, but for us, this will not matter, because I will look for the probabilities in other places.

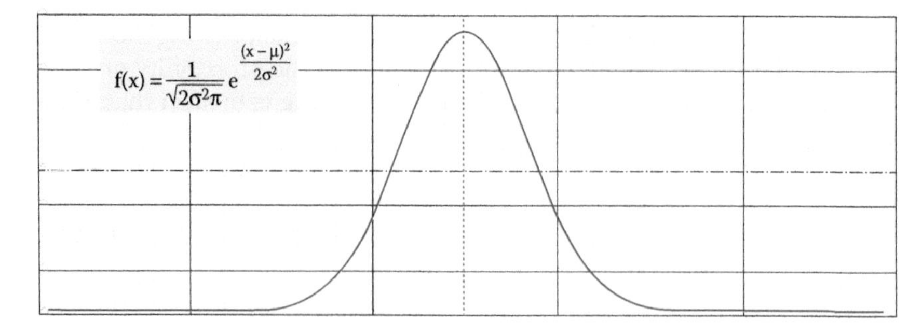

FIGURE 2.5
Formula and diagram for the Normal (Gaussian) distribution.

2.9 Combinatorial Analysis

To deal with probabilities of different interdependencies, I need to be familiar with variations, permutations, and combinations.

Variations, permutations, and combinations are used to calculate the number of elements in the sets that are connected with different rules. They are important for calculating different probabilities due to the necessity to know the number of all possible outcomes of a particular event. I can use Table 2.2 to show the simple difference between variations, permutations, and combinations, but the more important point is to use this table to determine which operation is connected with our calculations.

2.9.1 Variations (V)

Let us say that I would like to know how many two-digit numbers by using 4 digits can be written (there are four digits and I will use only two) and the order of elements is important (because it is making a difference between the numbers presented). Looking at Table 2.2, I determine that I should work with variations.

There are two possibilities. One possibility is to write two-digit numbers without repeating any of these four digits and the second possibility is to allow the repetition of any of these four digits. Mathematically, the first case is called "Variations without repeating," and the second one is called "Variations with repeating." The formulas for the calculations are:

$$V_n^k = \frac{n!}{(n-k)!} \text{ (Variations without repeating)}$$

$$\overline{V}_n^k = n^k \text{ (Variations with repeating)}$$

In both formulas (mentioned previously) n is the number of used digits (n = 4) and k is the number of digits in the created two-digit number (k = 2). For this case the results will be:

$$V_n^k = \frac{n!}{(n-k)!} = \frac{4!}{(4-2)!} = 12 \text{ (Variations without repeating)}$$

$$\overline{V}_n^k = n^k = 4^2 = 16 \text{ (Variations with repeating)}$$

TABLE 2.2

Determination of Operation

Elements of the Set Produced	Variations	Permutations	Combinations
Are all elements chosen?	NO	YES	NO
Is the order of elements important?	YES	YES	NO

As we can see, repeating increases the number of possible outcomes.

Variations with repeating are used when finding different ways of how a system made up from subsystems can fail. Let us say that I have a system that is made up from three subsystems (k = 3) that are connected in series. Having in mind that every subsystem could only have two states (normal and faulty), these are my elements (n = 2). So, I will have an overall set of states presented by the variations of three subsystems in two positions (normal or faulty) that are my elements.

Looking at Table 2.2, I need to calculate variations and I will use the formula for repeating*:

$$\overline{V}_2^3 = 2^3 = 8$$

In this case the number of variations with repeating helps me find a number of ways how a system built from three subsystems connected in series may fail.

If I mark the components that are functioning normally as S_1, S_2, and S_3 and situations when the components are faulty as S_1, S_2, and S_3, the full set (S) of possible situations will be:

$$S = \{S_1S_2S_3, \underline{S_1}S_2S_3, S_1\underline{S_2}S_3, S_1S_2\underline{S_3}, \underline{S_1}\,\underline{S_2}S_3, S_1\underline{S_2}\,\underline{S_3}, \underline{S_1}S_2\underline{S_3}, \underline{S_1}\,\underline{S_2}\,\underline{S_3}\}$$

In other words, using the algebraic set I can write:

$$S = S_n \cup S_f$$

where S_n is the set of normal situation presented by:

$$S_n = \{S_1S_2S_3\}$$

and S_f is the set of faulty situations:

$$S_f = \{\underline{S_1}S_2S_3, S_1\underline{S_2}S_3, S_1S_2\underline{S_3}, \underline{S_1}\,\underline{S_2}S_3, S_1\underline{S_2}\,\underline{S_3}, \underline{S_1}S_2\underline{S_3}, \underline{S_1}\,\underline{S_2}\,\underline{S_3}\}$$

With S_f we can see all of the situations where the full set (S) can fail and you can see that it will function normally only if every subsystem is functioning normally.

2.9.2 Permutations (Per)

If there is race of four cars and I would like to bet on the final result (sequence) of each car, it means that I need to guess which car will be first, which second,

* There is a possibility of two or three subsystems to be faulty at the same time, which means I am repeating the "faulty" element!

which third, and which fourth. The number of all possible situations (orders) regarding the sequences of the four cars is called a number of permutations* and it is given by the formula:

$$\text{Per}_{(4)} = 1 \bullet 2 \bullet 3 \bullet 4 = 4! = 24$$

If I assume that all cars and drivers are acting the same, then the probability to do the right guessing (only one of those 24 possibilities will be the right one!) is:

$$P = \frac{1}{\text{Per}_{(4)}} = \frac{1}{24} = 0.04166...$$

In percentage this means I have 4.17% chances of making the right guess (if all the cars and drivers are the same, which means that each of them has equal chances to be first!).

Let us say if from the four cars in the earlier example I want to guess which one will be first, then the probability of every car to be first is:

$$P_1 = \frac{1}{n} = \frac{1}{4} = 0.25$$

Now the probability of a car to end up second is calculated out of three cars (because one has already finished first) and is given by:

$$P_2 = \frac{1}{n-1} = \frac{1}{3} = 0.333...$$

Going further, I need to guess which car will be third, and my choice guess can be made from two cars (two are already first and second) and the probability for me making the right choice will be:

$$P_3 = \frac{1}{n-2} = \frac{1}{2} = 0.5$$

Now I do not have more choices, so the last car must be fourth and the probability for it to happen is 1 (100%):

$$P_4 = 1$$

A very important point to mention here is that I presented you with the degrees of freedom† in permutations, which is (n – 1). This means that from

* In accordance with Table 2.2, all elements are included and the order is important for the result!

† The degree of freedom is a number that shows the number of variable factors that need to be considered to achieve balance (the requested state) of the system. The rest of them are determined by the established connections with the already chosen factors.

four cars, I can try to guess the order of only three of them. The order of the last one (whichever car it is) will always be fourth.

2.9.3 Combinations (C)

In the case when I have to employ two candidates in my company out of seven candidates, in that situation not all elements (seven) are chosen (I need only two of them) and the order of them is not important. Table 2.2 determines the operation: Combination. Having in mind that I cannot choose one person for both positions, I have a situation of combination without repeating. In accordance with Table 2.2, the combinations and the number of all possible choices are given by the following formula:

$$C_n^k = \frac{n!}{k!(n-k)!}$$

For the above-mentioned case, n = 7 and k = 2, so the number of combinations will be:

$$C_7^2 = \frac{7!}{2!(7-2)!} = 21$$

It means that 7 people can be grouped in 2 in 21 ways.

If the elements can be repeated, then I have the formula for the number of combinations when they can repeat:

$$\bar{C}_n^k = \frac{(n+k-1)!}{k!(n-1)!}$$

Let us speak about the lotto.

In most countries, there are 39 numbers and I need to choose 7 of them. I can notice that not all elements are chosen (I need only 7 of them) and the order of choosing is not important. This means that I will use combinations calculating my chances of winning the lotto. Having in mind that elements (numbers!) cannot be repeated, I use the formula for combinations without repeating. The calculation of the combinations for this event (winning lotto) is given by:

$$C_n^k = \frac{n!}{k!(n-k)!} = \frac{39!}{7! \bullet 32!} = 15380937$$

So, my probability to win with one ticket is given by:

$$P_{Lotto} = \frac{1}{15380937}$$

In words, I need to play 15 million 380 thousands and 937 times lotto to have a reasonable chance to win it once! Please note that I said: ***To have a chance to win it***! This is very important because I can win the lotto with my first ticket or I may not win it even if I play three times more than 15,380,937. In general, I need to play 295,787 years and even then, I cannot be sure that I can win!

3

Statistics

3.1 Introduction

Statistics started before the eighteenth century and became mature enough at the beginning of the twentieth century. It can be defined as the science of processing (collecting, presenting, analyzing, and reasonably interpreting) data. Having in mind that there are different situations regarding the reasonable processing, there are two types of statistics.

The first one is the ***descriptive statistic***, which deals with general numbers (citizens in country, productions, etc.), and it is useful in the context of presenting a clearer and more understandable form of "a bunch of something." To fit into the "clear and understandable form," the data are usually presented with diagrams and some kind of markings.

The other type of statistic is the ***inferential statistic***.* It is applicable for a huge amount of data. Having in mind that these data create problems during their processing, there is a particular way of gathering and processing these. Instead of dealing with the whole data, I am using randomly chosen samples of them with the intention to obtain some general conclusion about the event (described by the whole data). The inferential statistics is divided into three disciplines: estimation, modeling relations, and hypothesis testing.

Even though inferential statistics is applicable to real life, it is used mainly in science, especially in areas where there are lots of data available. So, by using inferential statistics, I am trying to infer about a "general behavior" by calculations executed on a "group of samples."

As I have already said, inferential statistics is undertaking samples from one huge set of data (events, elements, numbers, members, measurements, etc.), and based on the statistical processing of these samples, it draws a conclusion for the full set. A simple example is the manufacturing of products that are produced with certain dimensions, but not all of them are checked. Workers randomly† choose particular samples from the production series

* Also known as Statistical Inference.

† A random selection is done only if I believe that all of the elements have the same characteristics and chances to be chosen.

(from the daily production). These randomly chosen samples are statistically processed. The results give the managers information about the overall situation of the products produced that day.

Both statistics are very important in industry because they are used for statistical process control (SPC), which is a terrific tool for the quality assessment of manufacturing processes. For the purposes of the Bowtie methodology (BM) and this book, I will only use the descriptive statistics because sampling in the area of quality and safety is very rare, and, additionally, I need all the data available for these calculations.

3.2 Statistics and Probability

When we deal with probability, we also deal with statistics! Probability is dealing with events which can occur accidentally and cannot be predicted (because they occur randomly). The main point about probability is that for its calculations, I need data. So, the real connection between probability and statistics is that statistics provides the data for probability. If you go back to Chapter 2, you can notice that probability differs from frequency by the number of available data needed to do the particular calculation. So, when the amount of data is big enough, I can speak about probability. More data provide better accuracy for probability calculations. If I use an infinite source of data, I can establish great knowledge for the particular experiment and I can easily calculate the probabilities of all possible outcomes regarding it. In all other cases, I deal with frequency.

Unfortunately, having enough data does not happen very often in reality....

To better explain the connection between statistics and probability, I will use the official data of the European Union (EU)* for 2014. In Table 3.1, we can see the data of fatalities by transport mode in the states of the EU for 2014 for both drivers and passengers.

TABLE 3.1

Fatalities in 2014 in EU States per Transport Mode as a Driver or a Passenger

Bus + Coach		Car + Taxi		Motorcycle	
Driver	Passenger	Driver	Passenger	Driver	Passenger
20	136	8120	3785	3626	222
Moped		**Agricultural Vehicle**		**Heavy Goods Vehicle**	
Driver	Passenger	Driver	Passenger	Driver	Passenger
702	38	140	25	498	96

* http://ec.europa.eu/transport/road_safety/sites/roadsafety/files/pdf/statistics/2014_transport_mode.pdf

TABLE 3.2

Statistical Data from Table 3.2

	Buss + Coach	Car + Taxi	Moped	Agricultural Vehicle	Heavy Goods Vehicle	Motorcycle
Driver (%)	0.15	61.96	5.36	1.07	3.80	27.67
Passenger (%)	3.16	87.98	0.88	0.58	2.23	5.16

Using the data from Table 3.1, I can make some statistical calculations which can be seen in Table 3.2. There I present the percentage of fatalities which are calculated using the number of fatalities for drivers (13,106) and passengers (4,302) (or in total: 17,408 fatalities).

Data presented in Table 3.2 are in percentage and if they should be "translated" into the "language" of probability, I can say that the probability of dying in an accident* using different types of transport means in countries in the EU is highest if you are in a car or in a taxi. Looking at the numbers, you can see that buses and coaches are the safest road traffic vehicles for drivers because the probability of dying there as a driver is just 0.15%. But if you are looking for the safest route as a passenger, then the best way is to use a moped (probability of only 0.88%).

Your chances of dying as a driver in car or in a taxi can be expressed as the probability of 61.96%, while dying as a passenger in a car or in a taxi is expressed as the probability of 87.98%. In other words, your chances of surviving a traffic accident if you are driving a car or a taxi are 38.04%, while the chances to survive as a passenger are 22.02%.

I can also express the statistical data in Table 3.2 by the following probability statement: During all accidents in the EU (based on their data for 2014), the probability of dying of in a car or in a taxi for a driver is 61.96%, whereas that of a passenger is 87.98%.

3.3 Mean and Standard Deviation

In Chapter 2, I explained the set as part of the probability theory. In statistics, I will speak about sets in the form of data. Having in mind that the mathematical definition of a set is an aggregation of all elements, members or participants in something under consideration and I use numbers to determine all of these points, sets are fully applicable in statistics. So, I can say that when the elements of a certain set can be expressed with numbers, then that set can also be treated as a set for statistical purposes. Usually, I put these numbers in a table or in a matrix and do not mention the term "set."

* By definition, an accident is a safety event where there are casualties or catastrophic damage of assets or environment. In this example the word accident is used due to the fatalities.

But the connections that exist for the elements (numbers) in the set actually have the same meaning as elements for set in probability. The universal set in statistics is known as ***population***.

The most important statistical calculations about the population or a sample of data chosen from population are: mean (μ), variance (σ^2), and standard deviation (σ). The mean is an arithmetical mean of a set of numbers and the formula* for calculating μ is:

$$\mu = \frac{\sum_{i=1}^{n} N_i}{n}$$

An example to help us explain the mean can be the point of balance when we put weights on their places on both sides of a scale.

Variance is given by the formula:

$$\sigma^2 = \frac{\sum_{i=1}^{n} (\mu - N_i)^2}{n}$$

The important point to understand here is that I use a square in the formula. The reason for this is that if the variability is expressed solely by the difference of elements from the mean value without it being squared, the result would be 0 as shown in the following:

$$\sum_{i=1}^{n} (\mu - N_i) = 0$$

You can check it. Some of the differences in the sums will be positive and the others negative, so, added together, they will give 0. By squaring when calculating the variance, we make the negative values of the calculations inside the sum positive.

Actually, the real measure for the variability (difference) of the numbers under consideration compared to their arithmetical mean is standard deviation (σ). The formula for calculating σ is:

$$\sigma = \sqrt{\sigma^2} = \sqrt{\frac{\sum_{i=1}^{n} (\mu - N_i)^2}{n}}$$

* Please note that this is a formula for descriptive statistics where I use all data to calculate values. In inferential statistics for the denominator, you can find n − 1 instead of n. The reason for that is that, in inferential statistics, you use samples of all data to produce an understanding of the population and that result is (if I use n) smaller than the mean of all data. So, to compensate this difference, I use n − 1 as the denominator instead n. In that context, if I use n − 1 then μ will become $\bar{x}$ (estimation of mean of population calculated by sample of data) and σ will become s (estimation of standard deviation).

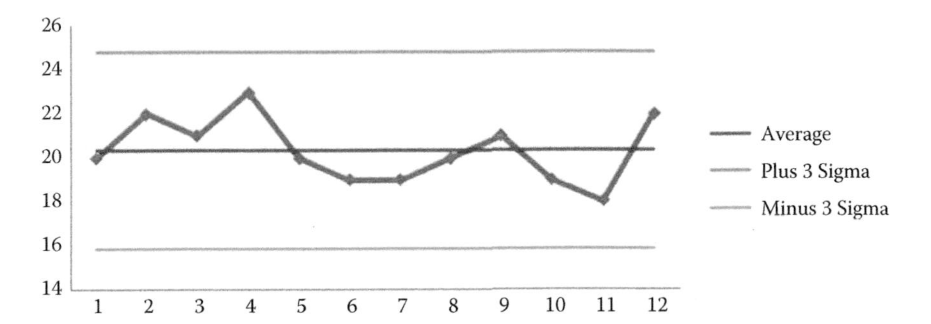

FIGURE 3.1
Example of descriptive statistics diagrams where mean (μ) and ±3σ (sigma) are marked. The blue line is connecting particular measurements, helping to see the variability of data.

In all formulas, N is the number describing the element, i is the serial number of the element, and n is the total number of elements. The element can be a measurement, data, number, result from calculation, and so on. Please note that the bigger the N is, the more accurate the calculations are.

In Figure 3.1, we have a simple example of descriptive statistics for the set of 12 measurements consisting of the following values: 20, 22, 21, 23, 20, 19, 19, 20, 21, 19, 18, and 22 with a mean $\mu = 20.33$ and a standard deviation of $\sigma = 1.43$.

The variance and the standard deviation are always positive, even though mathematically, the square root of the variance (which is actually the standard deviation) produces two solutions (one with + and one with −).

There are a plenty of software packages (commercial and free) for statistical calculations, but the cheapest way to calculate statistical data is to make tables in Microsoft Excel and put the data inside along with the previous formulas.

If I speak about processes of which their standard deviation in the SPC is big, I speak about unstable processes. For risk management, this can be understood as a series of events without a strong connection. Sometimes having a big standard deviation can be a sign of bad measurement, bad equipment, bad operator, and so on.

There is another point connected with standard deviation: the area of the graphic limited by $\mu \pm \sigma$ will contain 68.27% of all elements (measurements) and $\mu \pm 3\sigma$ will contain 99.73% of all elements (measurements). So, dealing with values which fall inside these limits gives us confidence with the processes (events, measurements, etc.) under considerations. In this context, the mean (μ) is a measurement for accuracy and the standard deviation (σ) is a measurement for precision.

There are also other values which can be calculated for statistical purposes, but they are not significant to our cases of safety.

3.4 Inferential Statistics and Safety

I use the statistical data to calculate the probabilities of events and I use these probabilities to make predictions.[*] Even though inferential statistics is used extensively in science, its use in safety is not good due to the context[†] of safety investigations (how you approach the safety investigations).

Let us explain why.

As I already mentioned, the inferential statistics is statistics where you take a sample of a full set which is presented by a big amount of data (population) and by investigating that sample, you infer about the characteristics of the whole set. For example, in the manufacturing industry where I am (hypothetically speaking) producing 100 products, I do not test them all. I am just sampling 10 of them and I am statistically processing only these 10. Based on the inferential statistical investigation for these 10 products, I decide about the characteristics of the whole set of 100 products produced today. If I notice that 9 of my samples are OK, I will say that 90% of the products manufactured today are OK. Statistic theory says that if my 10 products are randomly chosen, then their characteristics will represent all 100 products.

In inferential statistics, I calculate the estimated mean ($\bar{x}$) and estimated standard deviation (s) and they describe my whole set of elements. So, if I have 10 out of 100 products which produce an estimated mean value of (let us say) 1.5 mm and an estimated standard deviation of 0.2 mm, I should dedicate these values to the overall set which consists of 100 products.

But let us say that, between these 100 products, all of them are very close by value to $\bar{x} \pm 3s$. If I take one element (product) of this set and change it with another element which is not even close to the estimated standard deviation, let us say 7.5 mm (5 times bigger!), this product will behave to all other elements of the set as a Black Swan event. This situation is shown in Table 3.3.

So, the main question is: What will happen?

The answer is: Nothing and everything!

Look at the values in the table. For normal products, the samples' mean and standard deviations are almost the same as the mean and standard deviation of the set. But in the other part of table (the set with a Black Swan!), the situation is quite different. If a Black Swan is present in the samples (10 elements), it will change the value of the standard deviation a lot (in this case, 40%). Looking at the other situation, if the Black Swan is not present in the samples, then it will not be noticed at all because the values of the mean and the standard deviation are almost the same as the set without a Black Swan product.

---

[*] Maybe "prediction" is a strong word to use here. A more convenient one would be "expectation" because in reality we cannot predict, but when we have the particular probability of something, we can expect it. I am using "prediction" because it is a word which is far more familiar to people when speaking about probability.

[†] More on context in the next chapter!

TABLE 3.3

Normal Products and Products with One Black Swan

Sampling 10 out of 100	Normal Products		Products with One Black Swan	
	Samples (10)	**Full set (100)**	**... in Samples (10)**	**...in Full Set (100)**
Mean (μ)	1.5 mm	≈1.5 mm	2.1 mm	≈1.58 mm
Standard deviation (σ)	0.2 mm	≈0.2 mm	3.01 mm	≈0.2 mm

Note: This table is produced by me by using an Excel sheet and by simulating calculations on 100 virtual products. You can do it yourself too and you will see that the results will be almost the same.

In safety, we deal with rare events which are actually the outliers in our normal operations. Looking at this example with the Black Swan, you can notice that inferential statistics is just averaging or producing very wrong values. Let me emphasize one more time that in this particular example, if the Black Swan event is in the samples, then the information about the whole set of events will be extremely wrong, and if the Black Swan event is in the rest of elements, it will go unnoticed.

So, do not use inferential statistics in these calculations and pay attention to these rare events because they are actually what we are looking for.

3.5 Context of Statistical Investigation

Let us see a simple example: statistics about the matches of a football team in the last year.

If the team played 19 matches and won 10, draw 3, and lost 6, I can calculate the percentages of matches won, draw, and lost with the following formulas:

$$P_W(\%) = \frac{10}{19} \cdot 100 = 52.63\%,\ P_L(\%) = \frac{6}{19} \cdot 100 = 31.58\%,\ P_D(\%) = \frac{3}{19} \cdot 100 = 15.79\%$$

These calculations are not probabilities, because the set is very small (just 19). So, I can treat them as frequencies.

This is only for this year. If I want to compare the calculations of the matches in this year with the previous years, I need additional data which I will get from the results of the other years. And if I need to calculate the "chances of winning" in the next match, I will find the mean of all "winning" percentages in the previous years. The result will actually be a frequency, but nevertheless, I can use it to "predict" the outcome of the next match.

But this is pure mathematics and descriptive statistics. Inferential and descriptive statistics require the context of the investigation, so in this case, it is not enough to just calculate the frequencies of winnings and use them to calculate the expectation of winning the next match. I simply need more data different than the results (won, draw, and lost) to be more precise in my calculations.

Let us say that the team changed the manager this year, so he will probably change the style of playing. Regarding the new situation, obviously, the previous matches have nothing to do anymore because the subjects which were used to calculate probability (frequency) are no longer the same. Both my data and calculations are scientifically (mathematically) correct, but the conclusion based on these data will be false. So, statistics is extremely dependent on the context of the calculations! Be careful not to forget this!

It happens very often (intentionally or unintentionally) that people make mistakes because they use wrong contexts for calculations, so the interpretation of the result is wrong. To explain the context in statistics, I will use a joke: in almost every Balkan country, there is Sarma,* a very popular meal made when you boil leaves of cabbage already rolled around grinded meat. So if 10 people are eating meat and 10 other are eating cabbage, then statistically in mean, 20 of them are eating Sarma. This is absurd, but it proves my point that the context of calculation is of the utmost importance!

Let us go further.

If you go on the Internet and write in any browser "Funny Statistics," you will see about 50 million results. Try to open any one of them and you will see some funny things about statistics. These "funny things" are coming from the fact that if you do not take notice of the context, then the results are just for fun. There is an extremely good website named 22 Words† (http://twentytwowords.com) which I found. On this site, there is a page named "Funny Graphs" where some correlations between completely unrelated statistically analyzed events are presented. In statistics, correlation is used as a measure of how two or more sets of data (variables) change together (depend on each other).

If you open the page, you will see incredible similarities between the shape of the graphs presenting divorces in Maine (USA) and consumption of margarine in the USA per capita. Or age of Miss America and murders by steam, hot vapors, and hot objects in the USA. Even though there are extraordinary similarities with these graphs, they are totally independent and no connection can be established between them except the strange (and not-causal) similarities of the graphs between them.

I also remember another joke about statistics. When I was young, I heard a definition of a statistician which was presented in a political context: "If someone is lying for his own benefit, he is a liar! But if someone is lying for the benefit of the Government, he is a statistician!" In this case, I do not speak

* I think it is a Hungarian dish, but it does not matter: It is delicious!

† You can find it on Facebook!

about lying, but about using statistical data in the wrong context which produces wrong conclusion.

So, be careful when you are dealing with statistics:

1. Do not misinterpret the results by using wrong context.
2. Never forget that statistics is used to produce an approximation of a real situation!

The difference between science and fun in statistics is just a thin line which is very easy to cross.

4

Boolean Algebra

4.1 Introduction

Logic has been established as the science of thought, reasoning, and thinking, since the age of Ancient Greece. Aristotle was one of the first to dedicate his work to the logical thinking. In the middle of the nineteenth century, the British mathematician George Boole published a book with long name, but known today only as "The Laws of Thought." He was the first mathematician who tried to establish a systematic way in dealing with principles of logic and correct reasoning. That is the reason why the mathematical area of dealing with the quantification of logic operations is named (in his honor) Boolean algebra.

The Boolean algebra is actually a deductive mathematical system known as the algebra of logic and reasoning, and it had a dramatic influence to the computer development in the twentieth century. Even today, all processors use the Boolean algebra to conduct logic operations and calculate mathematical expressions.

Having in mind that Boole quantified the sentences (expressions) by their value (are they "true" or "false"), he used the binary numerical system to calculate their combinations. Later, this was accepted by computer engineers because the "true" could be expressed as 1 (current is flowing or voltage is present) and "false" could be expressed as 0 (current is not flowing or voltage is not present). The contribution of Boole is also important for the connection of probability with logic. It was a normal development of logic, using the probability of previous events to calculate the probability of future events. So, the Boolean algebra deals with logic and probability, and it uses the binary numerical system.

Actually, using 0 and 1 is not so simple. As I have mentioned earlier, the situation "1" (current is flowing) and "0" (current is not flowing) can be counted as switches. Switch in position ON means 1 (current is flowing) and switch in position OFF means 0 (current is not flowing). But again I can take also the opposite: 0 can be "current is flowing" and 1 can be "current is not flowing." This is known as negative logic, and it is used pretty often in electronic digital communications. The main point is that I am representing every bit of

TABLE 4.1

Symbols for Same Operations in Different Scientific Areas

Operation	Probability	Mathematics (Sets)	Logic (Boolean)	Engineering (Binary)
Union of A and B	A or B	$A \cup B$	$A \vee B$	$A + B$
Intersection of A and B	A and B	$A \cap B$	$A \wedge B$	$A \bullet B$ or AB
Complement of A	not A or A^C	A' or $\bar{A}$	not A or $-A$	A' or $\bar{A}$

information by two situations or states: true or false, 1 or 0, low or high, left or right, and so on.

It is important to mention here that there is no standardization between the use of the same symbols for operations used in different sciences (mathematics, logics, and engineering). Even in mathematics, there are different symbols than that in probability. Table 4.1 shows the variety of symbols for using the same operations in different sciences.

In this book, I will use engineering operations in accordance with Table 4.1. Note that symbols presented in Table 4.1 are not absolute. In different books you can find different symbols. I am choosing these because, in my humble opinion, they are most common. It means that if I have a probability of an event A and a probability of an event B, the probability of any of them to happen will be the sum of both probabilities*:

$$P(A + B) = P(A) + P(B)$$

Also, the probability that both of them (A and B) will happen at the same time is given as the product of the individual probabilities†:

$$P(A \bullet B) = P(A) \bullet P(B)$$

Looking at Table 4.1, I can notice that there is an equivalence between all these operations, which means that the laws presented in Table 2.1 are also applicable for the Boolean algebra.

4.2 Boolean Symbols for Engineering

Operations in the Boolean algebra are presented by the combination of particular symbols and their corresponding "Truth tables." "Truth tables" are tables of logic operations which give a particular "output" (true or false) for a particular combination of "inputs" (true or false).

* If A and B are independent events!

† If A and B are independent events!

TABLE 4.2

FTA Symbols (Gates) and "Truth Tables" for Them

AND				OR			
	A	B	A • B		A	B	A + B
	0	0	0		0	0	0
	0	1	0		0	1	1
	1	0	0		1	0	1
	1	1	1		1	1	1
XOR				**NOT**			
	A	B	Exclusive (A⊕B)		A	Not A	A
	0	0	0		0	1	0
	0	1	1		1	0	1
	1	0	1				
	1	1	0				
NAND				**NOR**			
	A	B	Not (A • B)		A	B	Not (A + B)
	0	0	1		0	0	1
	0	1	1		0	1	0
	1	0	1		1	0	0
	1	1	0		1	1	0

The symbols used in the Boolean algebra are the same symbols used in digital electronics. Actually, digital electronics is "engineering" the circuits which are fulfilling the mathematical and/or logical applications of the Boolean algebra. The symbols in Table 4.2 are called "Gates" in digital electronics.

Table 4.2 presents the symbols and "Truth tables" for any of them. I will use them in Fault Tree Analysis (FTA) to show the interdependence of causes and events.

Gates are used in safety and quality to explain the working of a system. For our purpose (safety and quality), I will use gates when building the diagrams to show the interdependence of the causes for the main event. Having in mind that these diagrams can be realized by electronic digital circuits, I will use the aforementioned gates to build a circuit for every such diagram. FTA can actually be executed by such electronic circuits built by gates.

The AND gate is used when I need both events (A and B) to happen at the same time to trigger the main event. The OR gate is used when I need only one of the events (A or B) to happen in order for the main event to happen. Of course, if both of the events (A and B) which are connected by an OR gate happen, the main event will also happen (Table 4.2). The XOR is used when I need only one of the events (A or B) to happen in order for the main event to happen. With the XOR gate, if both events happen at the same time, then the main event will not happen. This is the only difference between the OR and XOR gates. The NOT gate is used when the event must not happen for the main event to happen. The NAND gate is used only when both events happen in order for the main event not to happen. The NOR gate is used when both events happen (at the same time) for the main event not to happen.

It is important to mention here that the Truth tables for the gates in Table 4.2 are valid only if both of the events (A and B) are totally independent. If there is a dependency between them, then situation is quite different.

Let us remind ourselves of something from Section 2.5 (Basics of Probability). The probability of the main event to happen if any of three independent events happen is:

$$P(A,B,C) = P(A) + P(B) + P(C)$$

This can be presented with an OR gate where events A, B, and C are the inputs. But if they are dependent, then the formula will be:

$$P(A,B,C) = P(A) + P(B) + P(C) - P(A \cup B) - P(A \cup C) - P(B \cup C) + P(A \cup B \cup C)$$

So, this situation is not very simple. But there is a trick!

Please do not forget that the systems are designed to function normally! So, the malfunctioning does not happen very often, and speaking about its probability, the malfunctions have a very low probability of happening. Mathematically speaking, I am dealing with small numbers which will produce small mistakes if I miscalculate.

Let us see what happens if the events are dependent, but I use the formula for independent events (instead of the formula for dependent events). Looking at both previous formulas, I can see that the result of the formula for independent events will be bigger than that for dependent events. If you compare the formulas, this will be very clear:

$$P(A,B,C) = \boxed{P(A) + P(B) + P(C)} - \boxed{[P(A \cup B) - P(A \cup C) - P(B \cup C)]} + P(A \cup B \cup C)$$

The "colored" formula aforementioned is the formula for dependent events, and looking at it, I can see that the green rectangle is the formula for independent events. I can notice that I am subtracting the dependent probabilities (they are always bigger than 0[*] and are presented with the red rectangle) from the formula of independent events. So, this means that the probability of a few independent events happening is bigger than that of a few dependent events happening. This also means that if I choose to deal with the probabilities as events which are independent, then I use a more conservative approach or in other words I am safer if I use the formula for independent events than that for dependent probabilities!

How is this possible?

Having in mind that I use the FTA to calculate the probability (risk!) of an unwanted event (incident or accident), the results from using the formula

[*] Probabilities cannot be negative!

for independent events will give me a higher probability (of risk!) than the one for dependent events. It means that later, during the Event Tree Analysis (ETA), I will need to eliminate or mitigate the bigger risk (the worst case scenario). So, if my elimination or mitigation for this particular combination of independent events is good, it will be even better to eliminate or mitigate the risk calculated from the dependent events. By using this conservative approach, I will have more confidence (I will be safer!).

That which is mentioned earlier applies for the AND gate too. With the AND gate, I calculate the probabilities of the events if all of them need to happen in the same time.

If you look at Table 4.2, you can notice that there is an exclusive OR gate (XOR, $A \oplus B$), where the Truth table is almost the same as the one for the OR gate. The only difference is that when $A = B = 1$, the XOR gate will produce 0. This means that the main event will happen only (and only) if one of the input events is present. If both input events are present or absent, then the main event will not happen. This is valid only for independent events.

For dependent events, we have the following formula:

$$P(A,B)_{XOR} = P(A) + P(B) - 2P(A \cap B)$$

Again, using the XOR gate for independent events is better than using it for dependent events. Anyway, both A and B events are very rare, so having them happened in the same time is highly improbable for them to happen in the same time. The XOR gate is not of much use in the FTA, but you can never know whether you may need it.

In general, I recommend always calculating the probability of events as if they were independent. Even though I may be making a mistake using the formulas for independent events instead formulas for dependent events, I will be safer!

4.3 Boolean Functions

A simple example of a Boolean function is given in Figure 4.1.

If I analyze the combinations of different events as causes for the main event, this means that the different events (causes) are variable parameters in these combinations. In the Boolean algebra, dealing with the combinations of the variables is known as ***Boolean functions***. Mathematically, I can express the Boolean function as $f(x_1, x_2, \ldots, x_n)$. For the purpose of this book, all these variables $(x_1, x_2, \ldots, x_n)$ are events with a particular probability of happening. So, executing the Boolean function using these probabilities, I can calculate the value of the Boolean function (overall probability of the main event). These Boolean functions can be presented in three

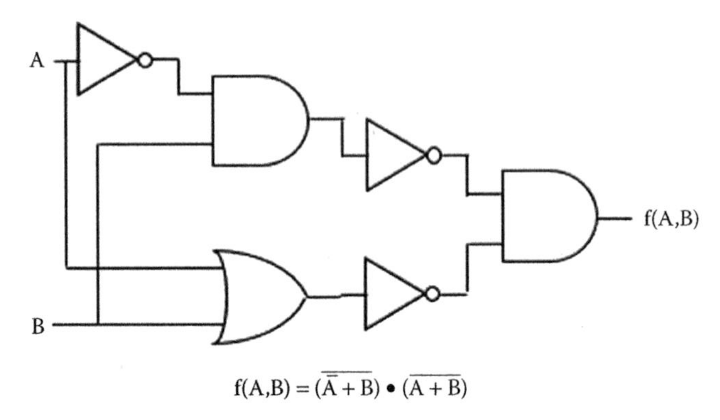

$$f(A,B) = (\overline{\bar{A} + B}) \bullet (\overline{A + B})$$

FIGURE 4.1
Boolean function presented graphically (above) and empirically (below).

TABLE 4.3
Boolean Laws Presented as Engineering Operations

Commutative laws	$A + B = B + A$	$A \bullet B = B \bullet A$	
Associative laws	$(A + B) + C = A + (B + C)$	$(A \bullet B) \bullet C = A \bullet B \bullet C)$	
Distributive laws	$(A + B) \bullet C = (A \bullet C) + (B \bullet C)$		
	$(A \bullet B) + C = (A + C) \bullet (B + C)$		
Idempotent laws	$A \bullet A = A$	$A + A = A$	
Law of absorption	$A \bullet (A + B) = A$	$A + A \bullet B = A$	
Complementation[a]	$A + \bar{A} = 1$	$A \bullet \bar{A} = 0$	$\overline{\overline{(A)}} = A$
De Morgan's laws	$\overline{A + B} = \bar{A} \bullet \bar{B}$	$\overline{A \bullet B} = \bar{A} + \bar{B}$	
Operation with 0 and 1[a]	$1 \bullet A = A$	$0 \bullet A = 0$	$\bar{0} = 1$
	$0 + A = A$	$1 + A = 1$	$\bar{1} = 0$
Combination of operations	$A + \bar{A} \bullet B = A + B$		
	$\bar{A} \bullet (A + \bar{B}) = \bar{A} \bullet \bar{B} = \overline{A + B}$		

[a] Please note that in Boolean and set expressions, I can use Ø and β which are actually 0 and 1 in engineering expressions.

ways: empirically (with a formula), using tables (known as Truth tables), and graphically (with a diagram).

The Bowtie methodology is dealing with causal safety.* This means that our analysis will show us different events (or combinations of events) which are causing our main event. These combinations can be presented as Boolean functions, and there are particular rules on how to calculate the values of these functions. In Table 4.3, the Boolean laws for calculating the Boolean functions are presented.

* Causal quality, causal reliability, and so on.

The important point that you need to notice within the laws in Table 4.3 is their dualism. Almost all of them are presented with two formulas which are similar and you can easily transfer from one formula to the other. To do that you need to change the operations (+ with · or vice versa) in the first formula, and then you get the second formula. In addition, for operations with Ø and β, you need to also swap the Ø and β values in the expressions. Of course, this does not mean that both of the formulas will calculate the same things. It only means that you can easily jump from one formula to another or in other words generate another expression for the same function. It can be seen if you look De Morgan's laws:

$$\overline{A + B} = \overline{A} \bullet \overline{B} \qquad \overline{A \bullet B} = \overline{A} + \overline{B}$$

The first one (left) explains how to calculate the probability of the main event (which can happen if any of A and B is present) not to happen, if I know probabilities A and B not to happen. The second one (right) explains how to calculate the probability of the main event (which can happen if both A and B must be present at the same time) not to happen, if I know probabilities A and B not to happen.

The total number of Boolean functions for n variables can be calculated using this formula:

$$N = 2^{2^n}$$

where N is the number of Boolean functions and n is the number of variables at the input (number of the inputs).

Do not be confused by the formula! There are n inputs and each of them is either 0 or 1, which means that the number of variations with repeating is 2^n. All of these variations (with repeating) will produce an output which is either 0 or 1. In total, I have 2^n input variations (with repeating) which are producing an output 0 and additionally 2^n input variations (with repeating) which are producing an output 1. Using the formula from earlier, for two variables (A and B), I will have $2^4 = 16$ different functions, and for 3 variables (A, B, and C), I will have a total of $2^8 = 256$ different Boolean functions.

4.4 Canonical or Standardized Form of Boolean Functions

I will have a finite number of Boolean functions, but I will have an infinite number of possible logic expressions that are equivalent to this finite number of functions. To eliminate the confusion, engineers usually use the so-called Shannon's method for expanding Boolean functions into a ***canonical*** or ***standardized form*** of Boolean functions. There are several such forms, but

TABLE 4.4

Truth Table for Function $F = AB + \bar{A}C$

A	$\bar{A}$	B	C	$A \bullet B$	$\bar{A} \bullet C$	$A \bullet B + \bar{A} \bullet C$
0	1	0	0	0	0	0
1	0	0	0	0	0	0
0	1	1	0	0	0	0
1	0	1	0	1	0	1
0	1	0	1	0	1	1
1	0	0	1	0	0	0
0	1	1	1	0	1	1
1	0	1	1	1	0	1

I will explain only two of them. The first one is called ***sum of minterms***[*] and the second one is called ***product of maxterms***.[†] To explain both of them, I will use a simple example of the Truth table for one simple Boolean expression:

$$F = A \bullet B + \bar{A} \bullet C$$

The Truth table for this function is presented in Table 4.4.

Looking at Table 4.4, I will define minterms as products of values of the variables A, B, and C which make the value of the function to be 1. Each A, B, and C with value 1 is presented by A, B, and C, and each value of A, B, and C with value of 0 (in the row) is presented as complements of A, B, and C (with the above-mentioned line!).

And similarly, the maxterms are defined as products of sums of values of the variables A, B, and C which make the value of the function to be 0. Again, each A, B, and C with value 1 is presented by A, B, and C, and each value of A, B, and C with value of 0 (in the row) is presented as complements of A, B, and C (with the above-mentioned line!).

Last column of Table 4.5 shows minterms (blue background) and maxterms (orange background).[‡]

So, the sum of minterms[§] will be:

$$F = A \bullet B \bullet \bar{C} + \bar{A} \bullet \bar{B} \bullet C + \bar{A} \bullet B \bullet C + A \bullet B \bullet C$$

So, the product of maxterms[¶] will be:

$$F = (\bar{A} + \bar{B} + \bar{C}) \bullet (A + \bar{B} + \bar{C}) \bullet (\bar{A} + B + \bar{C}) \bullet (A + \bar{B} + C)$$

[*] You can find it in literature as "sum of products."

[†] You can find it in literature as "product of sums."

[‡] The sum of minterms is also known as a disjunctive normal form of a Boolean function and the product of maxterms is known as a conjunctive normal form of a Boolean function.

[§] The rectangular shape surrounding the formula is corresponding to the color in the table.

[¶] The rectangular shape surrounding the formula is corresponding to the color in the table.

TABLE 4.5

Minterms and Maxterms for Function $F = AB + \bar{A}C$

A	B	C	$A \bullet B$	$\bar{A} \bullet C$	$F = A \bullet B + \bar{A} \bullet C$	
0	0	0	0	0	0	$\bar{A} + \bar{B} + \bar{C}$
1	0	0	0	0	0	$A + \bar{B} + \bar{C}$
0	1	0	0	0	0	$\bar{A} + B + \bar{C}$
1	1	0	1	0	1	$A \bullet B \bullet \bar{C}$
0	0	1	0	1	1	$\bar{A} \bullet \bar{B} \bullet C$
1	0	1	0	0	0	$A + \bar{B} + C$
0	1	1	0	1	1	$\bar{A} \bullet B \bullet C$
1	1	1	1	0	1	$A \bullet B \bullet C$

If I am present a function of the main event using the canonical (standardized) form (in this case F), then the sum of minterms shows the states (combinations of events A, B, and C) which are causes of the main event (F = 1). From another side, the product of maxterms shows the states (combinations of events A, B, and C), which are not the causes of the main event (F = 0).

But what are the minterms and the maxterms in the real world?

These are the scenarios how things happen for different combinations of variables in the system. Actually, all minterms are the scenarios of how the considered event (outcome) will happen and maxterms are the scenarios of how the considered event (outcome) will not happen. Of course, I can arbitrarily dedicate 0 to normal operation and 1 to faulty (abnormal) operation (and vice versa), and this arbitrary dedication is connected to the subject of the assessment (success or failure*). In this book, I am looking for "fault trees," so the "faults" will be 1 and the "successes" will be 0. Whatever markings I am using, the method of describing these scenarios is by applying the Boolean algebra.

This explains a very important feature of the BM: it does not only calculate the overall risk of a particular main event, but it also describes the scenarios of how it may happen. I can use this to deal with the risks (eliminate or mitigate them) before the main event happens.

4.5 Simplifying Boolean Functions

As I have already mentioned, the logic of Boolean functions can be presented as an electronic circuit. Sometimes logic tells us that there are different circuits

* More on this can be found in Section 6.5 of this book.

A	B	$\bar{A}$	$\bar{B}$	$\bar{A} \bullet \bar{B} = X$
0	0	1	1	1
0	1	1	0	0
1	0	0	1	0
1	1	0	0	0

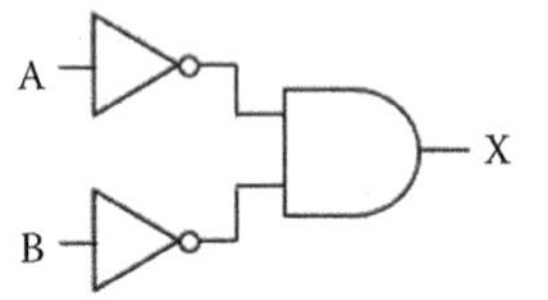

A	B	A + B	$\overline{A+B} = X$
0	0	0	1
0	1	1	0
1	0	1	0
1	1	1	0

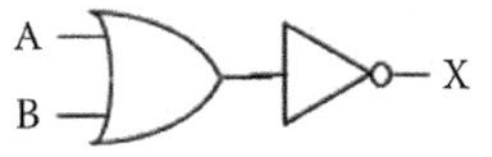

FIGURE 4.2
Example for equivalent circuits.

(Boolean functions) which present same outputs for same inputs. These are called ***equivalent functions***. There are plenty of situations when different circuits produce the same outputs when the inputs are the same and even though these circuits are not the same, we call them ***equivalent circuits***. In Figure 4.2, you can see two equivalent circuits with their Truth tables. You can notice that the circuits are quite different because they use different elements, but they are equivalent because the input is the same as the output. These two circuits actually present the first De Morgan's theorem from Table 4.3.

Comparing the Truth tables from Figure 4.3 with the Truth tables of Table 4.2, you can notice that the circuit presented in Figure 4.2 in the upper-right quadrant can be simplified with a NOR gate (lower-right quadrant) given in Table 4.2. Engineers noticed this, and it was the main reason why the NAND and NOR gates were introduced.

Simplification† can be achieved by using the Boolean laws from Table 4.3. Of course, instead of Ø and β, I will use 0 and 1, respectively. Simplifying a function to another equivalent function is important because digital electronics uses discrete components that produce circuits for a given function. Simplifying the function allows the designer to use fewer components, and this can reduce the cost of the design. In our case, we get a simplified picture of our system of faults which makes the analysis a lot simpler. It can be easily explained if you refer to Figure 4.3.

As you can notice, it is easier and cheaper to build a simplified version (orange color) than the canonical form of the function (blue color) presented by the sum of minterms. The benefit is much bigger for functions with more variables, like during the FTA when we have plenty of events causing the main event.

† Also known as Minimization or Optimization!

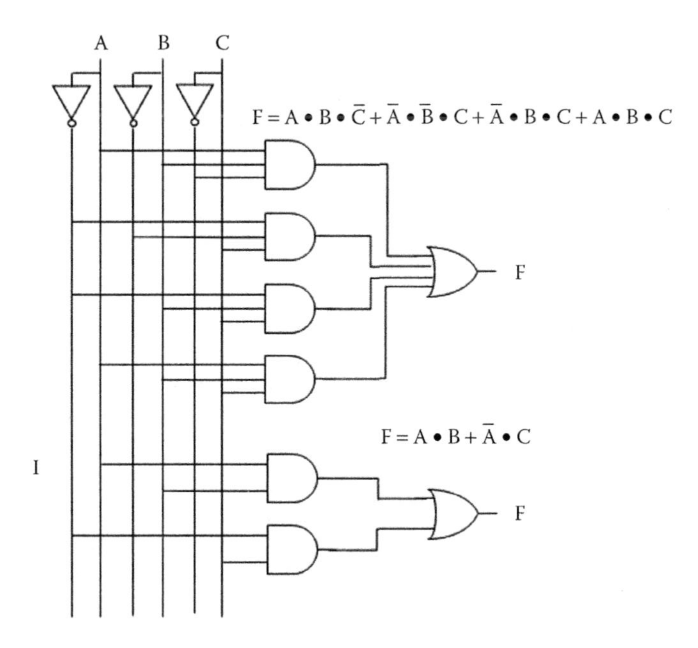

FIGURE 4.3
Comparison of the canonical form of the function (upper) and its simplification (lower).

During the simplification, it is easier to deal with the sum of minterms instead of the product of maxterms, so I will only use the sum of minterms in this book. Keep in mind that sometimes there are functions which produce Truth tables that have less 0s than 1s, and in these cases, it is (maybe) easier to work with the products of maxterms.

So, the overall process of simplification will be described by the following procedure:

1. Produce the Truth table for the function.
2. Use the Truth table to find the minterms (maxterms).
3. Construct the sum of minterms (product of maxterms).
4. Simplify the sum of minterms (product of maxterms) using the Boolean laws described in Table 4.3.

Let us see a simple example of the simplification of a function presented by the sum of minterms:

Original function $\rightarrow A \bullet B + A \bullet \overline{B} + \overline{A} \bullet B =$

by Distributive law $\rightarrow A \bullet (B + \overline{B}) + \overline{A} \bullet B =$

by Complementation $\rightarrow A \bullet 1 + \overline{A} \bullet B =$

by Operation with 0 and 1 $\rightarrow A + \overline{A} \bullet B =$

by Combinations of operations to **Simplified function** $\rightarrow A + B$

I will repeat again that both, in this example and for our purposes (finding the failure of a system), the canonical and simplified forms of the function present the combinations of events A and B which need to be fulfilled in order for the main event to happen. If I simplify the function presented by the product of maxterms, it means that the simplified function will present the combinations of events A and B which need to be fulfilled in order for the main event not to happen. And you need to be careful with that!

Simplification is extremely helpful during the FTA. Unfortunately, it is like a proving theorem in mathematics. There is no simple rule on how to do it and you need to improvise, to be innovative, to use your knowledge (in this case: the Boolean laws) and your experience to deal with it. Sometimes, realizing that a Boolean function cannot be further simplified can be a problem. For less than four variables, it is easy to notice, but if the number of variables is bigger, then the problem escalates. In general, when the minterms have nothing in common, it means that you have to stop the simplification.

There are other methods (Karnaugh maps, etc.) to deal with simplification, but they are more mathematically complex and thus I will not explain them in the book. Anyway, there is a lot of literature concerning the simplification of Boolean functions. The method presented in this chapter is enough for this book purposes.

4.6 Minimal Cut Set (MCS)

Simplification is important due to the ***minimum cut set*** (MCS). I can define the MCS as the smallest combination of variables which (when they happen at the same time) will produce 1 for the Boolean function. Translated for the safety area, it means that this is the smallest combination of events which can cause the main event. For the calculation of the MCS, I use the sum of minterms.

For the example presented in Figure 4.3, the MCSs are presented with this formula:

$$F = A \bullet B + \overline{A} \bullet C$$

It means that it is the combination of events A, B, and C which will produce my main event, and it can happen if A and B happen at the same time or if C happens and A does not. So here we have two MCSs: one is $A \bullet B$ and another is $\bar{A} \bullet C$.

The MCS can be of different weight and this property is called "order." Let us explain this in more detail:

$$F = A + \bar{A} \bullet B + A \bullet C + A \bullet B \bullet C$$

For the Boolean function as previously mentioned, there are four MCSs and these are: A (first-order MCS), $\bar{A} \bullet B$ and $A \bullet C$ (second-order MCSs), and $A \bullet B \bullet C$ (third-order MCS). In general, the bigger the order is, the harder it will be to eliminate or mitigate the risks in the FTA.

Why?

Because the aforementioned formula shows us that we need to deal with three events (A, B, and C) to achieve safety, and this is not always easy. But do not get discouraged as the things are not so bad. There is something that will help us very much. In the aforementioned formula, $A \bullet B \bullet C$ can be treated as a product of probabilities of events A, B, and C, and since all of them are less than 1, their product will be lower than individual probabilities. Or:

$$P(A) \bullet P(B) \bullet P(C) << P(A); P(B); P(C)$$

This means that I need to deal with a complex event made by a combination of three events A, B, and C which are happening at the same time, but the probability of this complex event to happen is very rare.

The first-order MCS is known as ***single-point failure*** (SPF). Remember that you must not allow having MCSs of first order!

Having SPF of first order means that only one single failure of one component could create the main event. This is not allowed. So, when the FTA shows that there is an SPF, it means that there is an existing subsystem (component, device, process) which is critical for the functioning of our system. I must immediately focus on transforming this SPF into an MCS of higher order. I can do this by adding another component (device, process) in parallel and a particular monitoring system which will switch the failed one with the good one (when it registers the failure of each of them).

For simple systems, the simplification of the Boolean functions provides a clear picture what might go wrong. In complex systems, looking for the MCS is an excellent tool to achieve an understanding of the functioning of the system and which problems you should prioritize. Nevertheless, this requires more knowledge and experience, not only in the area of Boolean algebra, but also in system configurations and system operations. So, whatever the Boolean operations are, the requirements for a good understanding of the systems are still valid.

5

Reliability

5.1 Introduction*

By definition, reliability is a special kind of probability that describes an event that must not happen (fault of equipment) for a particular period of time. In literature, in cases when this particular period of time (t) is determined, you can find it as ***probability of survival*** of the component (for the specified time t). So, we can define reliability as the probability of success (equipment not to fail) for a particular period of time.

There are many definitions of reliability, but my favorite is the one mentioned in the Military Handbook 338B (MIL HDBK 338B†). It states that reliability is the "probability that an item can perform its intended function for a specified interval under stated conditions." The main points in this definition are as follows: probability, specified interval, and stated conditions. I will not explain probability, but I will explain the meaning of specified interval and stated conditions.

The specified interval is a time period for which the equipment will not fail, and I would like to calculate the probability that (in this time period) equipment under consideration will not fail. For NASA, this time period should be time of the launching and the landing of rockets (as they are the most critical parts of every space mission), for airlines this should be time for the take-off and landing of the aircraft (as they are the most critical parts of every flight), and so on. The most important point is to remember that reliability is not to show the lifetime of the equipment.‡

Stated conditions are the situations that are directly associated with this equipment. It means: When the equipment is working in particular environment and under particular conditions, the operation is stated as normal. So, for reliability, I need to calculate the probability that this normal situation will continue for a particular period of time (specified interval!).

* In this chapter, I present only the basics of reliability, or in other words, the fundamentals that are enough for the calculations necessary for the use of reliability in BM calculations.

† Issued on October 1, 1998, by the Department of Defense of USA.

‡ In Figure 5.1, there is a diagram in which the life of the equipment (product) is explained!

It is very important to have enough data to calculate this probability. The amount of data available will show you if you are working with frequencies or probabilities. For particularly risky industries, there are enough data available from the regulatory bodies. Risky industries are strongly regulated by laws, rules, and standards, so the regulatory bodies have an obligation to oversee the company's performances and gather data regarding safety.

There are a plenty of regulatory documents that deal with reliability analysis using the FTA and you can find most of them on the Internet. The most famous are the NUREG-0492, the famous "Fault Tree Handbook,"* issued by US Nuclear Regulatory Commission. Another one is MIL HDBK 338B, "Electronic Reliability Design Handbook"† issued by the Department of Defense in USA. Two other very useful books for electronic equipment are the MIL HDBK 217F (Reliability Prediction of Electronic Equipment) and the Telcordia‡ SR 332 (Reliability Prediction Procedure for Electronic Equipment). In addition, there are other books that deal with mechanical parts of products.

Using such handbooks, companies that produce equipment provide the calculated reliability for their products presented in the form of MTBF. But, companies that bought this equipment should prove these values during the usage of this equipment. So, they need to keep record of all of the failures of the installed equipment and to use it to calculate the MTBF on their site. In industry, data gathered using the product in practice in a reasonable period of 4 years§ should be used to calculate the value of the MTBF that must be nearly the same as the one predicted from the manufacturer. If these two values are not similar, something is wrong with the calculations or with the measurements.

Reliability is mostly used in industry. There, it is connected with the equipment installed in the factories. Nevertheless, it is fully applicable to the products sold to customers. In this book, I will speak about both, even though in this chapter I will be speaking mainly about equipment.

5.2 Basics of Reliability

Reliability can be calculated by the formula:

$$R(t) = e^{-\frac{t}{MTBF}} = e^{-\lambda t}$$

* Used mostly for teaching purposes!

† Guidance material only!

‡ The company called Bellcore, who changed their name to Telcordia (somewhere you can find it as Telecordia), made a revision in MIL HDBK 217 for telecommunication purposes and published it as SR 332.

§ In aviation, for navigational equipment, this period is 2 years. In general, longer period equals better results when calculating reliability in reality!

where:
t is time (specified interval!)
MTBF* is the Mean Time Between Failure

The MTBF is calculated as:

$$\text{MTBF} = \frac{\text{AOT}}{\text{n}}$$

where:
AOT is the Actual Operating Time (time during which I am using the equipment)
n is number of failures

The MTBF is connected with the Failure Rate (λ), which is expressed by the formula:

$$\lambda = \frac{1}{\text{MTBF}} = \frac{\text{n}}{\text{AOT}}$$

To have this probability expressed in percentage, simply multiply by 100. In general, the longer the MTBF is, the better is the reliability, so I can use the MTBF as measurement (unit) for reliability. But you should understand that MTBF is a statistical value and because of this it carries with itself all of the good and bad things of statistics.

To clarify reliability, I will use a simple example. Let us assume that specified interval t = MTBF, so the formula will give me the following result:

$$R(t) = \text{e}^{-\frac{t}{\text{MTBF}}} = \text{e}^{-\frac{\text{MTBF}}{\text{MTBF}}} = \text{e}^{-1} = 0.37$$

This result can have different meanings. The first one is (if I speak about all products produced in one batch) the fact that after the time period equal of the MTBF, only 37% of the products from the series will be operational. The second one is (if I am speaking for one product) that after a period of time equal to MTBF, the probability that this product will work is only 37%. This means that after specified interval of MTBF, I need to monitor my product more often.

To put this in context, let us make a few other calculations: for 1, 2, and 3 years of equipment which has an MTBF of 30,000 h. Results are given in Table 5.1.

* In literature, the term MTBF is used for repairable items and MTTF (Mean Time To Failure) is used for nonrepairable items! So, for repairable items I use the MTBF and for nonrepairable items I use the MTTF in the reliability formula.

TABLE 5.1

The Reliability That an Equipment with MTBF = 30,000 h Will Have after 1, 2, and 3 Years

Specified Interval (t)	1 year (8,760 h)	2 years (17,520 h)	3 years (26,280 h)
Reliability	0.75 (75%)	0.56 (56%)	0.41 (41%)

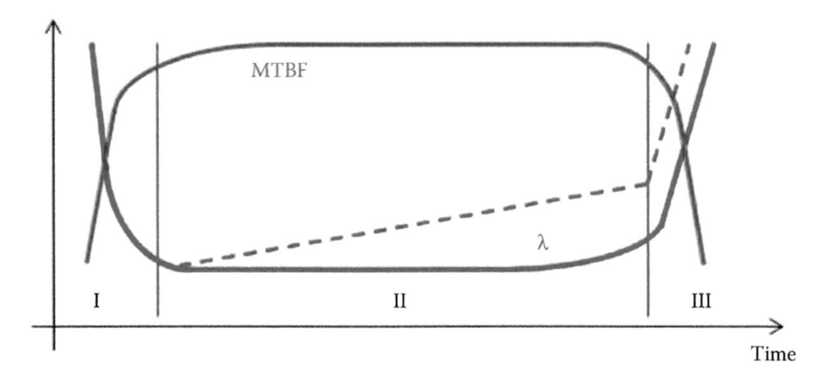

FIGURE 5.1
Reliability expressed by Bathtub curve for MTBF (blue curve) and λ (red curve) of the equipment during its life. (Full red curve is for electronic equipment and the dashed line is for mechanical equipment. Mechanical equipment has an increasing line due to wearing that is not so evident for electronic equipment.)

I can notice that as time passes, reliability decreases. This is because we are dealing with an exponential function with a negative exponent in the calculations of reliability.

In Figure 5.1, the lifetime of electronic equipment (product) is described. As you can see, in the beginning of the operation of equipment (region I*), there are plenty of faults due to the time needed for the equipment to adjust to the environment and due to the time needed for the operational and maintenance personnel to adjust to the latter, you can see that the reliability and the MTBF are entering the mature zone (region II in Figure 5.1) where both values are constant. When the MTBF tends to increase (λ is going up and reliability is going down) as presented in region III, we should think about changing the equipment due to increased need for maintenance. Region III is the time when the number of faults is drastically increasing and the maintenance is becoming too often and too expensive.

* In some literature, region I is part of the aging processes in the manufacturing facilities before the equipment (product) is released in the market and there is nothing wrong with that! Anyway, the same diagram applies when the equipment is installed in a factory. Installation is actually the start of the lifetime of the equipment and decommissioning is the end of it. Here I am speaking about the useful part of the lifetime of the equipment, after the process of installation.

Anyway, there is a need to monitor the reliability all the time because it is a valuable source of information on how the equipment behaves, how the probability of failure changes, and when it is time to buy new equipment.

There are a few other points that need to be mentioned here. The reliability will be improved if the AOT* is big and it can happen only if the NOT† is small! Nonoperating time is the time that the maintenance personnel will spend to fix the faulty equipment. This is expressed as MTTR (mean time to repair). Again, this is statistical value and assumed to be not more than 30 min for industrial equipment.

5.3 Reliability of Complex Systems

The systems that I refer to consist of many components‡ and to calculate the probability of the overall system I need to take the reliability of every component into consideration. There are three situations of connecting the components: series, parallel, and combined (series and parallel).

For the series connection of the components (Figure 5.2), I have a situation when the output of one component is connected to the input of the next one and so on. For an example, this is a connection that I use when I need to amplify some signal, so I put a few amplifiers to achieve the proper amplification without distortion.

To keep the serial system functioning, I need all of the components to be OK. This means that for series systems, the formula is:

$$R_S = R_A \bullet R_B \bullet \cdots\cdots R_N = e^{-\frac{t}{MTBF_A}} \bullet e^{-\frac{t}{MTBF_B}} \bullet \cdots\cdots \bullet e^{-\frac{t}{MTBF_N}} = e^{-\frac{t}{MTBF_S}}$$

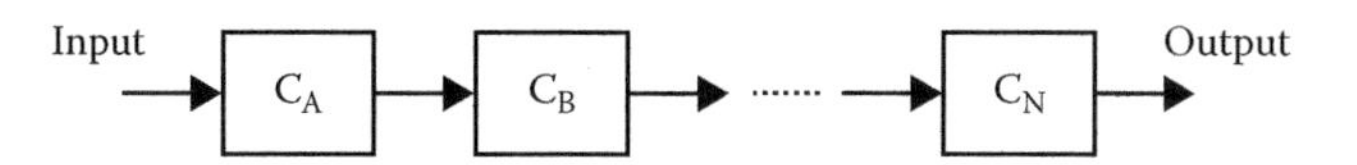

FIGURE 5.2
Series connection of components.

* AOT is acronym for Actual Operating Time. The AOT can be explained as the time of using the equipment. Let us say, I have a car for 3 years, but I drive it only 2 h per day. So, the AOT will be 2 (hours) multiplied by 3 (years) multiplied by 365 (days). More on this in paragraph 6.3.

† NOT is Non-Operating Time. In the context of the explanation of AOT, it is the time when I am not driving my car.

‡ I am dealing here with components, but this is fully applicable to subsystems!

Regarding the MTBF, I will have:

$$\frac{1}{MTBF_S}=\frac{1}{MTBF_A}+\frac{1}{MTBF_B}+\cdots\cdots+\frac{1}{MTBF_N}$$

From the formula, I can see that the $MTBF_S$ is going to be lower than any other individual MTBF (than any individual λ) in the formula.

Regarding the λ (failure rate), I will have:

$$\lambda_S=\lambda_A+\lambda_B+\cdots+\lambda_N$$

Formula for λ says the opposite: λ is going to be bigger than any individual λ in the formula. So, connecting the components in series decreases the reliability. But, not everything is about reliability. There is another problem with series connection: If one component fails, the whole system fails.

Parallel connection is presented in Figure 5.3. This connection is usually used to achieve redundancy of the equipment: I use two transmitters, connected in parallel (both connected to a same microphone and antenna), so if one fails, the others continue to broadcast. Usually, there is a circuit for automatically transferring the microphone and the antenna to the good transmitters if one fails.

Using components in parallel connections is a way of improving reliability. In addition, with parallel connection, failure of a component will not endanger the whole system. It will continue to operate, but the effectiveness and efficiency will be altered.

Even though N parallel components are shown in Figure 5.3, it is very rare to use more than two components in a system. The reason for that is that more redundancy means more costs, more complexity, and more instability.

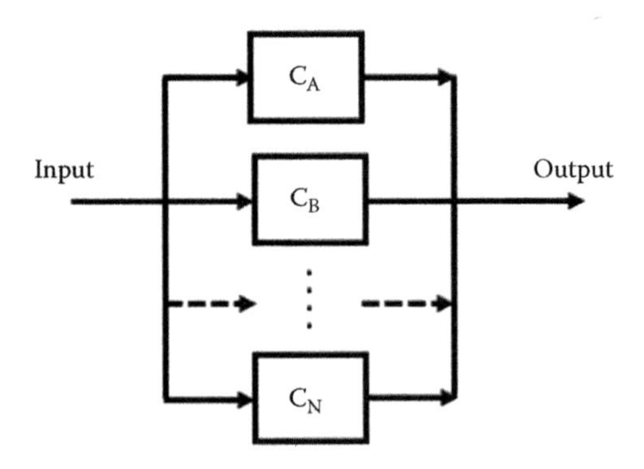

FIGURE 5.3
Parallel connection of components.

Putting two components in parallel in the systems increases the costs for the price of both (the second component and the monitoring and controlling device that switches to the second component when the first one is faulty).

In addition, all complex systems need to be controlled and maintained by humans, and "human reliability" is not the best. It means that with the parallel configuration in the system, I am not only implementing an additional technical component but also a human component. So, I am increasing the complexity of a system and complex systems are prone to more instability due to the doubling of the variables that influence normal operation of the system.

So, beside the pros of reliability, there are also cons caused by the complexity of a system. In my opinion, the benefits (better reliability) are much bigger than losses (bigger complexity), so I strongly recommend using parallel systems for improving the reliability of operations.

Please note that formulas for systems with parallel components are obtained if you use a double negation of reliability. The reason for that is that it makes the calculations much easier.

The negation of reliability is expressed with this formula:

$$P(n) = 1 - P(f)$$

In general, if I have N nonidentical components ($R_1 \neq R_2 \neq \ldots \neq R_N$), the formula will be:

$$R_s = 1 \overset{2}{\downarrow}{-} \overbrace{(1-R_1)\bullet(1-R_2)\bullet\cdots\bullet(1-R_N)}^{1}$$

If I have N identical components ($R_1 = R_2 = \ldots = R$), then the formula will be:

$$R_s = 1 - (1-R)^N$$

The first negation(s) is presented with the expressions marked with 1 and by $(1 - R)^N$. This actually represents the negation of reliability of different components in the system under consideration. The second negation in the previous formula is expressed by the minus marked with 2. It means that the reliability of normal operations is presented through the full probability of the set minus the faulty operations.

I will use a different approach to show the consistency of the Boolean algebra in the case of reliability calculation. For this approach, I will start with the formula for two components, which have reliability values of R_A and R_B. The overall calculation will be:

$$\overbrace{R_S = R_A + R_B}^{1} = \overbrace{R_A + \overline{R_A} \bullet R_B}^{2} = \overbrace{R_A + (1-R_A) \bullet R_B}^{3} = \overbrace{R_A + R_B - R_A \bullet R_B}^{4}$$

To explain the result, I will use the reliabilities R_A and R_B and the Boolean operations from Table 4.3, which totally apply for the previous formula. It is normal to assume that the reliabilities of the components with parallel connection will add (**1**). But it is not very simple because B will only work if component A fails, which means this situation can be presented with (**2**) by applying the upper part of Combination of Operations from Table 4.3. Having that in mind, we continue with:

$$\overline{R_A} = 1 - R_A$$

I can write (**3**). Executing the multiplication, I am getting the part (**4**) from equation as a result. This means I can write the formula for the reliability of two parallel components as:

$$R_S = R_A + R_B - R_A \bullet R_B$$

You can get the same result using the formula:

$$R_s = 1 - (1 - R_a) \bullet (1 - R_b) \Rightarrow R_s = 1 - (1 - R_A - R_B + R_A \bullet R_B) = R_A + R_B - R_A \bullet R_B$$

Combined systems (series and parallel combinations of elements) can be solved using the formulas for reliability of both series and parallel systems. The operation will strongly depend on the failure of the component. If a series component fails, then the whole system fails. It is good if you can group the combinations into series and parallel configurations and calculate them separately. Later, you just calculate the overall reliability using the formulas for series or parallel combinations.

A simple combined system is presented in Figure 5.4. You can notice that the system is "broken into" three components and the reliability of each component can be calculated individually and be multiplied later. So, the formula for calculating the reliability of this combined system from Figure 5.4 will be:

$$R_S = R_A \bullet (R_B + R_C - R_B \bullet R_C) \bullet R_D$$

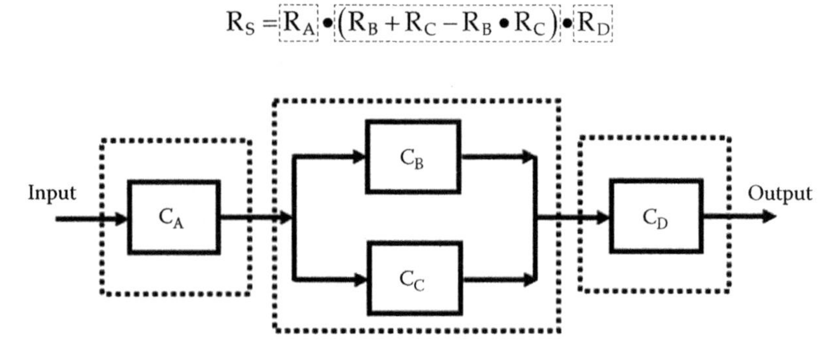

FIGURE 5.4
Combined (series–parallel) connection of components.

5.4 Reliability of Equipment and Reliability of Services

Reliability is connected with equipment, so let us speak about that a little bit.

If I have a critical system that needs to process the operation, then I need high reliability. This happens in aviation: Almost all of the CNS equipment in the aircraft and at the ground is doubled with the intention to increase the reliability of the equipment. It means that if one of the transmitters in the system on the ground fails, there is another one (back-up transmitter!) that will be "triggered" by an automatic monitoring device and the system will continue to transmit the signal. Aircraft in the flight will not notice change of the transmitters.

These transmitters are connected in parallel and the reliability with two transmitters will be bigger than simply having one transmitter. So, having the previous paragraph in mind, I can calculate the reliability for this "doubling" of the transmitters.

Let us assume that I have an ILS that is sending a navigational signal to an aircraft, and signal is used by aircraft to safely land on the runway. Both part of the ILS (the Glide path and the Localizer) have two transmitters connected automatically. Looking at each of them, I can assume that if one transmitter has a reliability of 15,000 h expressed as MTBF, then its reliability for 1 year time will be:

$$R(t) = e^{-\frac{t}{MTBF}} = e^{-\frac{8670}{15000}} = e^{-0.578} = 0.56 = 56\%$$

Putting one more transmitter in parallel, assuming that both are of the same type and same reliability (0.56 or 56%), the reliability of the combination will be:

$$R_S = R_A + R_A - R_A \bullet R_A = 2R_A - R_A^2 = 2 \bullet 0.56 - 0.56^2 = 1.12 - 0.3136 = 0.8064 \approx 0.81$$

To be honest, I must say that in the previous formula I should also calculate the reliability of the automatic circuit that will monitor the two transmitters, switching to the other if one of them fails. But, for the sake of reality, I will decrease the calculated reliability a little bit and the context of explanation will remain the same.

So, you can notice that the reliability was 0.56 and now it is 0.81, which is an improvement of almost 45%. Expressing the new reliability through the MTBF, I will get:

$$R(t) = e^{-\frac{t}{MTBF}} \Rightarrow MTBF = -\frac{t}{\ln R(t)} = -\frac{8670}{\ln(0.81)} = 41144.4455 \approx 41144$$

Looking at the result, you can notice that the new MTBF is 41,144 h, which is 2.75 times more than the MTBF of a single transmitter (which was 15,000 h).

But, putting one more transmitter did not improve the reliability of the equipment! Both transmitters will still have the same reliability (0.56) as other parts of the equipment. That which I improved is actually the reliability of the services offered by these transmitters. In aviation, these two transmitters are actually offering navigational service through the signal radiated by them. I know that a lot of reliability engineers will not fully agree with me on transferring reliability from the equipment (products) to the services, but I found this necessary to be expressed here for the sake of reality.

The increase in reliability is evident, but this new MTBF actually calculates the time between two outages of the system and a lot of manufacturing companies state that this is actually MTBO (Mean Time Between Outages), which differs from MTBF. This means that the outage in our case is the loss of the signal (signal is not available for the aircraft) caused by the failure of both transmitters.*

So, in general the MTBF applies to single equipment and the MTBO applies to services offered by parallel combinations of equipment.

5.5 Reliability of Humans and Organizations

Reliability is connected with equipment, but I just explained that the services also have reliability and the measure for it is the MTBO.

But how can I speak and calculate the reliability of humans and organizations?

This question is more connected to the latest development of Safety Management. A long time ago, the safety science established that humans are a weak link in all safety-related events. In aviation, 80% of all causes for incidents and accidents are human errors. This percentage in road traffic is even higher. But, investigating human factors responsible for these unintentional human errors brings a new "player to the game": The organization of the work. Bad management in a company is the biggest reason for human errors. So, yes, I can speak about the reliability of humans, and especially of the organizations, but I cannot calculate it in the same manner as that of the equipment. This means that I need to find another way to calculate it, and this new way has to come from the probability calculations based on the history of previous events and human involvement in them.

Improving human reliability can be achieved by the so-called Poka-Yoke methods, which are actually error proofing methods.

Speaking about the reliability of organization, I consider the safety culture built inside the organizations, but still there is a lack of measurement

* Word "outage" is used very much in all industries to express suspension of operation for particular time.

methods for that. The organizational reliability can increase only through the capabilities, understanding, and personality of managers. For the time being, there is no systematic solution.

5.6 Using Reliability for Probability Calculations

As I mentioned before, reliability is part of the specification of the products in the industry and there are a lot of books written about it. Reliability is well known in industry and it is one of the main quality specifications. High reliability means that the product will last more time. In aviation, it is a part of the regulatory requirements, but such a regulation does not apply to other industry. In aviation, it must be calculated because every fault of the equipment on the ground or in the aircraft can have serious safety consequences.

There are handbooks in which all the data for the MTBF calculations are mentioned, so the engineers can calculate the reliability for every product (equipment). These calculations are time consuming, so I will not go into detail to explain them. As I said, it is not easy, but it is obligatory that the reliability is expressed in MTBF. In fact, this is a regulatory requirement in all risky industries! When the equipment is installed, there is a procedure that needs to prove that the manufacturer's calculations about the reliability are true.

Here I would like to emphasize something else, which should produce more awareness when using reliability in safety calculations.

Reliability is the probability that the equipment will function normally for a particular period of time, but I am actually interested in the probability that the equipment WILL NOT function normally for a particular period of time. It means that for safety purposes, the reliability is applicable by its negation. So, the probability that the facility will NOT be operative (faulty!) will be given by the formula*:

$$P_{(faulty)} = \overline{R(t)} = 1 - R(t) = 1 - e^{-\lambda t} = 1 - e^{-\frac{t}{MTBF}}$$

For overall systems with known reliability, the previous formula can be used to find the probability that the system will not function normally (will be faulty!). If I do not know the total reliability of a complex system, then I can use the same formula for complex systems (series, parallel, and combined). But I need to apply the De Morgan's laws on them (Table 4.3).

* The important approximation that I can do is for $\lambda t < 0.001$, then $P_{(faulty)} = \lambda t$, but this is applicable only for simple manual calculations.

So, for systems made of three components connected in series, the reliability is presented with this formula:

$$R_S = R_A \bullet R_B \bullet R_C$$

But, the probability that the system will not function normally (negation of R_S) is given by the formula:

$$\overline{R_S} = \overline{R_A \bullet R_B \bullet R_C} = \overline{R_A} + \overline{R_B} + \overline{R_C}$$

In such a case, the total probability that the system will be faulty (due to De Morgan's laws) will be the sum of the individual probabilities of components (that the components will be faulty).

There is a similar situation with complex systems made by parallel components. If the system has two parallel components, then the formula for reliability is:

$$R_S = R_A + R_B$$

The probability of having a failure in the system (negation of R_S) after the implementation of De Morgan's law is given by the formula:

$$\overline{R_S} = \overline{R_A + R_B} = \overline{R_A} \bullet \overline{R_B}$$

Combined system will use a combination of the formulas for both series and parallel connections.

6

System

6.1 Introduction

In this book, I speak about practical implications of the BM, which is used in QMS and SMS, so it is essential to explain the definition of a system in these two management systems. QMS and SMS are management systems in which the managing of quality and safety is done systematically by building procedures that define the interactions between humans and equipment. A system in QMS and SMS (for purposes of this book) is defined as an aggregation of humans, equipment, and procedures. Actually, the system is shaped by the written procedures (system and operational) that are setting up rules on how the humans need to behave and interact with the equipment to produce the wanted outcome (safe products or safe services with a particular level of quality).

An important point to mention here is that even the humans and the equipment can be treated independently as a system. Humans in companies are grouped into departments and units and each of them is actually a part of the company treated as an operational system. Similarly, the equipment can be treated as a system too, because it is made of at least two independent parts: mechanical and electrical. In addition, the electrical part of the equipment can be divided into hardware and software. These, together, form a technical system.

Fortunately, for the purposes of BM, there is no need to make a difference between a management system, an operational system, and a technical system (equipment, machines, instrumentations, etc.). The BM is applicable to everything. So, for purposes of this book, a ***system**** can be anything (humans, equipment, material, activity, process, etc.) that is involved in a particular operation, and if some of those things are missing or not functioning properly, the operation will not be conducted normally.

Let us assume that the system is made up of smaller parts called subsystems. It is highly logical to apply the BM for all of these subsystems individually

* This is the definition that I usually use when speaking about QMS and SMS. Someone might not like it, but this definition offers solution to the problems, so I prefer it.

and correlate them later to see a clear picture of the overall system. Anyway, even these definitions are relative. Sometimes, a particular subsystem (consisting of small parts) can be built as a smaller system of these parts (which can be also treated as subsystems). Again, there is no difference if the BM is conducted on the system or on the subsystem.

I conduct the BM for particular Main Events, which happen inside the limits of a particular system. The Main Events are just fractions of the processes inside the system and by connecting them I can produce an overall picture of the safety of the system. When used as safety analysis of a system, the BM is requesting an excellent understanding of the overall functioning of the system in advance, but it takes care only of the Main Events that are related to safety (those that can produce safety consequences).

6.2 Critical Systems

In safety literature, the term ***critical system*** is repeated very often. It is interesting that there are a lot of definitions explaining this term. Some of them are highly dependent on the context of the operation (safety, quality, science, business, finance, etc.), but most of them deal with the fact that critical systems contribute most to the failure of an operation (activity, process, etc.), and in that way causes incidents, accidents, or potential losses for the company.

There is another term that is used in literature, and it is also dedicated to the criticality of the systems—***dependability***. It is defined as a characteristic of a system that describes how humans trust the system. Of course, most of the critical systems have low dependability. Every safety or quality method for system analysis provides some degree of trust for the system under investigation. Of course, if I am looking for a wider context of the usability of a system, then dependability is not the only characteristic of a critical system. I can say that there are many other factors that are also included, and I cannot rely solely on the outputs of the methods used for the quality and safety analysis. For example, maybe a system will provide the required quality and safety (high dependability), but it will be too expensive to manufacture. This means that this system will not bring any economic benefit to a company: The trust in quality and safety will be high, but the trust in bringing a profit makes it unsustainable. Anyway, the BM will provide important results regarding dependability, but it does not mean that this will be enough to say that the system is fulfilling the expectations. In general, good systems do not have a big dependability. As the dependability increases, the costs of the functioning and maintenance of the system also increase.

The criticality of the system or its subsystems can be measured by the frequency of how often they contribute in triggering some "bad things to happen." Of course, the systems prone to failures are more critical because they are contributing more to the abnormal functioning of the operations. Another measure for criticality can be the scope of the effect of failure of this system. If it affects more than one part of the operation, its criticality is bigger.

The next measurement for criticality can be connected to the applications inside the system that take care of monitoring and controlling the system. If their contribution to the system failure is too big, then it creates more criticality (low dependability!). The type of monitoring and control also affects the criticality. For the sake of safety, quality, and economy, it is always better to provide more electronic circuits for the monitoring and control than humans. Do not forget: Humans have extremely low reliability and integrity compared to equipment! So, the criticality goes up if the system is controlled by humans and goes down if the system is controlled by electronic equipment.

Anyway, in this book, I will use the terms system and critical system as the same. There are two reasons for that. The first is that the word ***critical*** is very subjective. For some companies, losing few hundred thousand USD is critical, and for some, losing few million USD is critical. The company that finds it critical to lose few millions USD will not care about losing few hundred thousand USD. The second reason is that I am using in the book some of the methods for analysis of risk (for the events, operations, activities,...) which I determined as critical for satisfying the requirements for ordinary systems. So, the things that apply to ordinary systems will also apply to critical systems.

6.3 Functioning of the System

I will establish a system where there is a need to deal with a particular operation. Some operations need to be conducted in a management system, some in an operational system, and some in a technical system (equipment, machines, instrumentations, etc.). Subjects that are involved in an operation are usually humans who are following particular procedures and using particular equipment. I agree that a system can also consist of only equipment or only humans, but for the purposes of this book, it will not make any difference.

The operation, if you look it up in dictionaries for definition, is always connected with processes. Different dictionaries provide different definitions and there is nothing wrong about using the term "process" instead of the term "operation." I would not debate about this here, but I assume that an operation is a bigger entity than process, so I can say that one operation may have more processes inside. In that context, I would assume that the operation of risk assessment for a particular system consists of a few processes: Gathering data for the system, evaluating the data, building a structural tree as a representation of the system, analyzing the tree, determining severity and frequency, making a decision for the acceptance of the risk, preventive and corrective measures, and so on. Whichever it is, everything that applies for operations is also applicable for processes.

As I said, I use operations to finish some job. An operation needs time to produce an effect. If an operation is a success, it means that I have achieved my goal and I can say that the operation was normal. In the case when the goal is not achieved, the operation was maybe wrongly planned. But, this is not always the case. Operations can be designed to be good, but some of the subjects involved in the operation (conducted by a system!) could fail during its execution. So, to continue with this operation, I need to investigate what the problem was and solve it. Every operation has its normal time of operation and also abnormal time, when there is a failure of some of the elements inside. In such a case, I will stop the operation to solve the problem. Regarding the operations, I can define the time for operation T_{OP} as:

$$T_{OP} = AOT + NOT$$

where:

AOT is the Actual Operating Time
NOT is Non-Operating Time

AOT is the time during which the operation is conducted normally and it leads us to the expected effect. NOT is actually the time when there is some failure in the system, so the operation cannot be conducted normally, which means it brings us bad results (the goal is not achieved!). Another name for NOT is time for "incidental maintenance."

No one likes NOT. It prevents us from getting the desirable profit because in order to solve the problems that are producing the NOT, I need additional time and I take this time from the AOT. As the NOT increases, the AOT decreases due to the constant T_{OP}. In the case of a NOT in production process, my production process stops and I cannot produce my product and because of that I lose money. In addition, to fix the problem that caused NOT, I need to engage other resources (employees, tools, equipment, energy, etc.), which means I lose even more money.

Failures in the systems are the biggest reason for NOT, but they are not the only reason for it. Since no one likes NOT and failures are the biggest reason for it, there are good reasons to investigate the failures and implement measures that will decrease the NOT. These measures are known as preventive maintenance. Let me remind you that every 10,000 km (or every year) you bring your car to the workshop to change the oil, the filter for the oil, and so on. This is all preventive maintenance: I am taking a small part of my AOT and I am giving it to the NOT, with the intention to prevent longer NOTs (caused by failures) in the future. This affects the reliability of the whole system.

Figure 6.1 has a reliability curve (expressed by λ) of a piece of equipment. Let us assume this is my new car. Please note that this is just a simulation, which

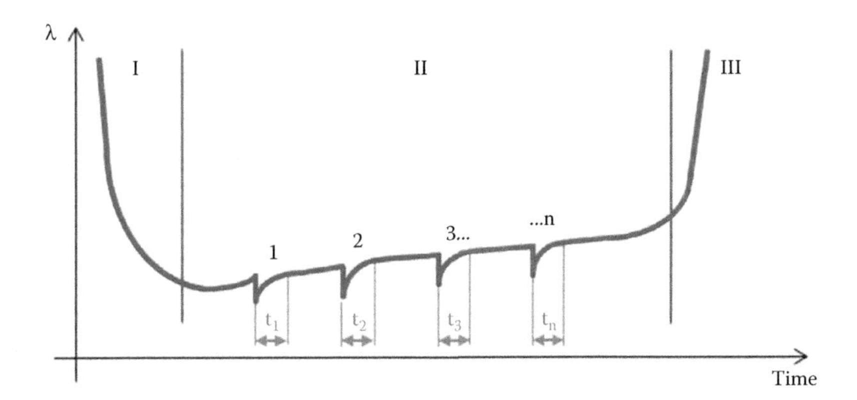

FIGURE 6.1
Changes in reliability (expressed by λ) due to preventive maintenance.

may differ for different types of cars. After buying a car, I should take care not to drive it more than the limits for (at least) the first 1,000 km. Different car companies propose different values, but the main point is that the car needs to adapt to the style of the driver and to the environment. Actually, cars are extensively tested during the design process and the aging of the produced cars is not conducted. So, at the beginning, my car will have smaller reliability. After 1,000 km, I should go to the workshop (position 1 in Figure 6.1), where the maintenance personnel will change the oil and filters and adjust the valves and screws where it is necessary, so my car will have an improved reliability, which is presented as the lower failure rate (λ) in Figure 6.1. After the first workshop visit, I should keep visiting it on a regular basis. This is presented by the numbers 2 and 3 in Figure 6.1. Of course, there will be a lot of visits, not only two. However, my car will not last forever, so there will come a time when even after my regular maintenance is done, my car's reliability will not go up the same value (situation 4). This happens simply because the preventive maintenance is not applicable to all car parts, so due to wearing, some of them need to be changed. But at a certain point in the future, I will realize that the maintenance is too expensive in terms of money and time, so I will probably sell the car to someone who does not have the money to buy a new nonused car, but has the money and time to deal with the maintenance. I will buy a new car and I will be at the beginning of diagram again (Figure 6.1).

For my calculations connected to the BM, I will neglect the NOT because of the maintenance. Simply, for the purpose of the BM, it does not matter. It is always scheduled and there is no failure that can be calculated. But do not forget, failures can sometimes happen due to bad maintenance too, so when looking for a cause (especially with the FTA), you must not forget that!

There is one more thing that can be noticed from the diagram. Look at the values of t_1, t_2, t_3, and so on. These are actually the time values that extend the AOT and as such provide a better value for the MTBF (smaller λ) and improve the overall reliability of my car.

6.4 System Characteristics

The first thing I need to do is to acquire knowledge about the system and ask for other people's experiences about its functioning. Knowledge can be found in books or manuals and experience can be gathered by speaking with people who have already dealt with this system. I cannot always have an excellent knowledge and understanding of the systems, so there is nothing wrong to consult somebody. Actually, I think that this is the best way to get familiar with the system—to speak with somebody who is already working with it!

One of the most important characteristics of the system is how big the system is. This is important because it will determine the external boundaries of the system. If I want to use the BM on a system of aviation navigation, it will consist of two segments: aircraft (equipment in aircraft) and ground (equipment which is installed on the ground). In between is the air (a medium for the transferring of the radio signals). This system is operated and maintained by humans (pilots and maintenance people) that are following particular procedures. The equipment in both the aircraft and the ground are paired with transmitters and receivers, but there is also an Inertial Navigation System (INS) in the aircraft that is autonomous (not paired with a ground segment). So, will I need to include the INS into my analysis because it functions quite differently than the paired navigational equipment? If I decide to deal with aircraft segments only, then I will assume that the signal sent by the ground segment is error-free and I will analyze only the equipment in the aircraft, the operations by the pilots, and the procedure used.

Figure 6.2 shows an overall system for navigation in aviation. It can be any ground-based navigation aid represented by the transmitters (blue) with ground-based monitoring (green), the air as a transmission media (black), and avionics (equipment in aircraft, red). The meaning of the blocks is as follows: M1/M2 are monitors, TX1/TX2 are ground transmitters, DB is the

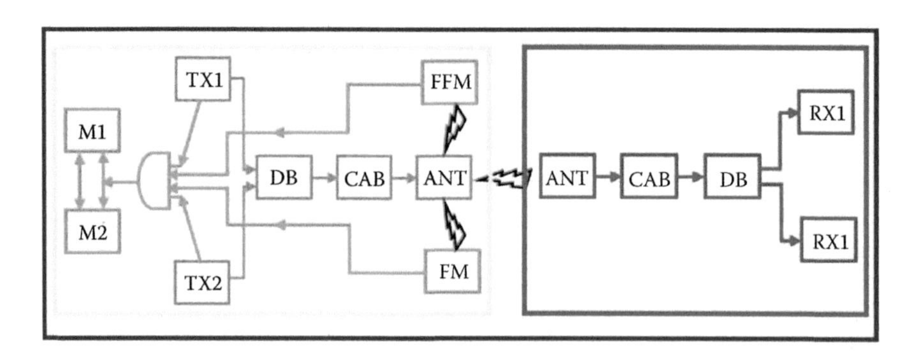

FIGURE 6.2
Diagram of the overall system for navigation in aviation.

distribution box, CAB are antennae cables, ANT are the antennae, and RX1/RX2 are receivers. The arrows represent the navigational and monitoring signals in the systems. Depending on the requirements or my wishes, I can assume a few of the external boundaries of the system. I can implement the BM only on the transmitters (blue), on the monitors (green), on the avionics (red), on the ground segment (yellow), and on the complete system (violet).

There are a few good things regarding the external boundaries of the system.

The first one is that I am building the system starting from the Main Event, which means that I am the guy that determines what is important and what is not. As a first step, it is OK for me to think that everything is important, but later (during analysis) I may see that some things can be neglected.

The second good thing that I need to do is to set the boundaries of the system in accordance to my perception of the problem I want to investigate. Let us say that I would like to see how the driver's fatigue is contributing to accidents, so I neglect the other factors if they are not connected with fatigue. So, the problem that I would like to investigate shapes the boundaries of our system.

The third important thing is that boundaries are flexible. Be open-minded: Whatever the boundaries, they can be changed during the analysis. So, adapt your perception of the system to the real needs. As you investigate the situation, maybe some new moments will show up, so you should feel free to change the system.

There is one more type of boundaries. These boundaries are inside the system and because of that they are called "internal boundaries." In literature, you can also find the term ***limit of resolution*** because these boundaries show us how detailed the analysis of the system will be. These boundaries are determined by answering the question: How deep do I want to go to find the cause of the Main Event? If there is a malfunction of some equipment in industry, it is probably caused by a fault in a subsystem or component. The smallest subsystems in the industry's equipment are the Line Replaceable Units (LRU). So, this means I need to find the faulty LRU and replace it. It will fix my problem and there is no need to continue. But if the same replaceable unit fails very often, then I will need to go further with my analysis to the smallest component inside the LRU and the analysis will stop when I determine the component or reason why the LRU is failing so often.* Searching for a smaller element instead of the LRU will actually make the execution of the BM more complex. But do not forget that even the internal boundaries are flexible and they can change during the analysis. Going back to Chapter 1 and Figure 1.2, I represent the internal boundaries of the system with the volume of events 11–20, which are marked as quaternary (IV).

Boundaries should be defined clearly, as they are very important for every analysis. In general, working with smaller internal and bigger external

* The component is maybe falling due to high current, which is submitted to LRU from the power supply. So, my failure is not in this LRU, but inside the power supply.

boundaries is always better, especially at the beginning of the execution of the BM. It gives a better picture about the situation in the system. Later, you may change it, but they must always be clearly stated in the BM report. Anyway, changing the boundaries must be triggered by the need of analysis itself, not by the results of the analysis. Importance of this statement can be presented with an example: A rounding of decimal numbers.

There are rules for rounding and it is very interesting that many of the experienced engineers make mistakes when rounding numbers. Let us mention the rules here because rounding is necessary in later stages of calculations.

If the number (which needs to be rounded one step up*) is:

- Bigger than 5, then I increment the number. It means that the rounding of 1.986 will result in 1.99.
- Smaller than 5, then I decrement the number. It means that rounding of 1.982 will result in 1.98.
- Equal to 5, then the situation is different. If the number before 5 is even, then I just delete the 5. This means that the rounding of 1.985 will result in 1.98. If the number before 5 is odd then I increment the previous number. This means that the rounding of 1.975 will result in 1.98.

Table 6.1 presents the mathematical rounding of the sequences of numbers.

I can see how different they are. The first one is a value of some measurement presented as a Result‡ and all the others are mathematical rounding of this Result presented by different resolutions of the instrument. I can see that the difference between the resolutions Integer and Result (4 decimals!) is 0.0575, which is actually 6.1%. Sometimes 6.1% can be negligible, but when dealing with big numbers, it is considerable. Anyway, if I change the internal boundaries of my analysis from 0.9425 to 1, then I am making a mistake and

TABLE 6.1

Mathematical Rounding of Sequence of Numbers

Sequence	**0.9425**	**0.942**	**0.94**	**0.9**	**1**
Resolution	Result	3 decimals	2 decimals	1 decimal	Integer
Error (±)	0	0.0005	0.0025	0.0425	0.0575
Error (%)	0	0.05	0.3	4.5	6.1

* "Rounding one step up" means that rounding a number with three decimal places will produce a number with two decimal places (or the rounding of 1.42 will produce 1.4). Going left with my decimals means that I am actually "climbing" the scale of numbers.

‡ To be scientifically correct, I should do (at least 10) series of measurements, then I should use these results to calculate the average (μ) and standard deviation (μ), and I state both of them (μ and σ) as the result of this measurement. So, for the purpose of the table, I use the average value (μ) as a result.

this mistake (of 6.1%) will produce a possibility that some of the probabilities expressed by those 4 decimals will disappear.

This rounding is very important for statistics also. There, you should take care of the rounding so you do not lose the significance of your calculations. With a lot of data, all similar to each other by value, this loss of significance is very much possible. So, the general rule is to do rounding after the calculations are done and only when you are sure that it will not change anything in the final result. On the other hand, keeping more decimal places gives you false assurance that you are precise, even though no information is hidden in this precision. In safety, we are dealing with low probabilities, so the rounding of numbers is extremely sensitive and should not be done before the final calculations.

6.5 Failure or Success

The newest developments in the area of safety are presented by the terms of Safety-I and Safety-II. Safety-I deals with faults where I try to investigate how the bad things happened ("what is going wrong"). This is part of a reactive approach to risk management where I use my knowledge (lessons learned!) to prevent "bad things" from happening again. Safety-I is causal and, by its ontology, reactive. Starting from the 1990s, with implementation of the systematic approach to Safety-I, method of investigating the risk evolves into proactive with intention to become predictive in the future.

Safety-II has different (I would say "opposite") approach. It does not investigate "what is going wrong," but instead it investigates how the system could achieve normal functioning ("what is going right"). The main point with Safety-II is to improve the normal functioning. The logic behind this is very clear: By putting efforts to make the systems work normally, I am actually decreasing the situations when the systems can fail. By using probability terminology, I can say that I am looking to increase the probability that the systems will function normally. This makes the probability of failures (abnormal operation) smaller. Concept of Safety-II is actually concept of Quality-1* because in the quality area I am doing the same thing: improving normal operations.

Roughly, Safety-I deals with failures and Safety-II deals with successes.

There is no scientific objection for Safety-II approach. My humble resistance toward this is connected with the practical realization of the overall idea. There is a beautiful example in the NUREG-0492 book (*Fault Tree Handbook*), which

* This concept is introduced in my previous book "Quality-1 is Safety-II: The Integration of Two Management Systems." This is actually ordinary quality used all over the industry today and in the past.

presents the Fault Tree and the Success Tree. Actually, these two complement each other and, as explained in the example, the Fault Tree is simpler than the Success Tree. The reason for that is that systems usually succeed much more than they fail.

A simple example for this is any kind of situation in the industry production. The production of products in industry by its ontology assumes that normal process will produce normal products. By definition, in the Statistical Process Control (SPC), a normal process* is a process adjusted to produce products that satisfy the specifications expressed by tolerances. Tolerances are limits for the variables of each process parameter. So, this way you have considerably more good products (success produced by normal functioning of the processes) compared with the number of scrap that are products resulting from abnormal functioning of the processes (failures).

But let us see how the scrap happens?

Figure 6.3 shows a diagram of tolerances for a particular product. Let us say that the product is good if the length is between 5.4 and 5.6 mm. Values of 5.4 and 5.6 are actually the limits of tolerances for this product and each value inside it could be a result of a different mode of variability inside the process. But I do care about that because each one of those variabilities will produce a success of my product. As I can notice, the number of the values inside the tolerance limits is infinite, which means that the number of successes can be infinite. But, there can be only two numbers of failures: 5.4 and 5.6 mm. If the measured value is smaller than 5.4 mm or bigger than 5.6 mm, it will produce scrap. Actually, the number of the points for scrap is also infinite, but for me, the only points of interest are the two points (5.4 and 5.6 mm) that present the limit of failure modes. So, success modes can be presented by many points and failure modes can be presented only by these two.

Some people say that they do not care about these successes that are inside the tolerance limits. This means that they are only interested if the product is inside or outside the tolerances, but in reality, it is not like that. There are plenty of companies that classify their products as Class I, Class II, Class III, and so on. This classification represents the distance of the "product" from the average value between the tolerance limits. In addition, I need to do

FIGURE 6.3
Tolerances of the length of a product.

* Using this definition, I could define abnormal process: It is a process that produces products that do not meet the specifications expressed by tolerances.

Measurement System Analysis (MSA) on the values of tolerances to see if the instruments I have for the calculation of Quality Control are good or not. So, successes are very important and failures do not always count. I read somewhere that a good manufacturing company would keep scrap levels at less than 5%, which means that the number of successes will be 95% (failures will be at 5%).

But there is another important thing with successes. If I made a mistake, then I am no longer working in the success tree. In that case, I am in an unknown area and there the probabilities of a second, third, and so on mistakes are 30–80 times higher. This is extremely valid for humans. The success tree assumes that humans are following the procedures on how to execute normal operations and that they are well trained for these procedures. So, if they make a simple mistake, they will end up in unknown areas where they must improvise. A good knowledge of the system may help during this improvisation, but Human Factors (fatigue, low concentration, panic, etc.) show us that this is not always the case. If this mistake happens, then I should consult the fault tree (where I have assumed all "bad situations"), which will put me in a better position to fix the mistake.

Anyway, the history of Safety-I has proven itself very successful, especially in aviation, so I will stick to it. There is no objection to build a Success Tree (dealing with Safety-II) instead of a Fault Tree (dealing with Safety-I), bearing in mind that they complement each other. The presentation of the Fault Tree is made up of the Boolean expressions (which are actually Boolean functions) and it is easy to produce complements of these functions. Simply, by complementing all the events and substituting the AND with OR gates and vice versa (De Morgan's laws!) in the MCS, I produce a minimal path set* (MPS). In this case, MPS is the smallest combination of variables which will produce 0 (success for our purposes) for the Boolean function. Translated for the purposes of safety, this means that this is the smallest combination of events that will not cause a faulty Main Event. It means that I am using the product of maxterms to calculate the MPS.

6.6 Faults and Failures

I will use FTA for functional safety. Functional safety† (FS) is an area of safety that deals with the safety of the products (machines, gadgets, appliances, etc.) or services (transport, medical, etc.) for different types of customers.

* Sometimes called ***minimal tie set*** (MTS), which is actually MPS.

† There is another type of safety called Occupational Health and Safety (OHS), which refers to taking care of safety in factories, where the products (machines, gadgets, appliances, etc.) are manufactured, or offices or companies, where services (transport, dinning, haircutting, etc.) are offered. Of course, there is no reason why FTA should be used in OHS.

In FS, all possible events that can cause an incident and accident originate from faults of the equipment, failures of humans, or failures of procedures.

As you have already noticed, I use the words ***fault*** and ***failure***.* Fault is mostly connected with equipment (things or machines) and failure is connected with operations (activities and processes). If I am driving my car with the intention to reach some place, not reaching this place is failure. A reason for that can be a car defect (fault!) or simply not knowing the right way which would lead me to this particular place (failure of human knowledge!). So, every fault will cause a failure, but not every failure is caused by faults. There are failures that are caused by the deficiency of humans or procedures in the operating of the system, which means that failures can also cause failures. Sometimes faults are caused by failures (bad training for the operating of the equipment or bad maintenance). In general, in this book I speak about failures of operations and faults of equipment.

Failures and faults can be categorized in three categories.

The first one consists of faults or failures that happen inside the environmental area of operation. The environmental area of operation for particular equipment is given as a specification of the environmental conditions necessary for the equipment to function normally.

Let us say if the car is produced and tested in Africa or the Middle East, then putting it in North Canada or Alaska does not guarantee its normal functioning. So, equipment is optimized for a particular environment. When I say "particular environment," I do not only mean parts of the world, but everything that is necessary for the equipment to stay in the limits of the specifications for normal operations. Under environmental specifications, I assume particular temperature, pressure, humidity, and so on. So, this is fault or failure that happens due to bad design, bad material, or bad use of equipment even though environment requirements are satisfied.

The second category is faults or failures that happen outside the specifications for normal operation. Simply, the equipment (and the humans) satisfies particular requirements for proper functioning and if these requirements (expressed as specifications) are not fulfilled, then the normal equipment is outside its limits and faults or failures can happen very easily. With humans, these "outside limits operations" result when operating under stress (tiredness, panic, etc.), which cause failures. For the machines, putting in a truck more weight than specified by specifications for use can cause a fault of the truck. Putting more weight in an aircraft will prevent take-off and may cause an accident.

The third category is just mistakes produced by humans, mostly unintentionally. Sometimes it is negligence, sometimes fatigue, sometimes wrong

* On the Internet, different dictionaries have different definitions. Especially in safety and quality areas, there are always different definitions. For the purposes of this book, I use the definitions presented in this paragraph.

training (lack of knowledge or skills), not following procedure, and so on. But such things happen extremely often and this is reason to worry.

Causes of third category of faults or failures are known as ***human factors***. In aviation, human factor errors are the main cause for 80% of all accidents. It does not mean that equipment fails just 20%. In aviation, there is a system of emergency procedures that are executed when there are equipment faults with the intention to maintain the normal operation of aircraft until it lands. But humans, very often, fail to follow even these procedures.

In general, there is need to be more conservative when calculating the probability of human errors than when calculating the probability of equipment faults. I will repeat again: Human reliability is very low compared with equipment reliability.

6.7 Effects, Modes, and Mechanisms

In literature, you might find the terms of failure effects, failure modes, and failure mechanisms. Those are not very important when building a Fault Tree, but during the analysis, they matter a lot. The analysis will need to establish interactions between the elements (events) in the Fault Tree and because of this you need to understand the differences between these three terms.

Failure effects are consequences of the happening of some event (failure). To distinguish between these effects and the effects of the Main Event, I will call them ***internal effects*** because they are internal part of FTA.

Internal effects can be found as an answer of the question: What are the consequences of the failure? Whatever happens, I cannot stop it because it had already happened, and so, I need to focus on the consequences. From a point of safety and in connection with the Main Event, dealing with these effects should be part of the Event Tree Analysis. But not really! The effect of every event in the Fault Tree is part of the Fault Tree Analysis, so these are effects, usually with fewer consequences than the Main Event. These effects are important because they are warnings that a certain situation is moving toward triggering a Main Event, and so I need to react to these events. Dealing with effects of primary (secondary, etc.) events will improve the probability that the Main Event will not happen. So, the control, elimination, and/or mitigation of the internal effects can be better if achieved during the FTA (Pre-Event activities!).

Failure modes are connected with knowledge collected from the previous events that triggered a Main Event. In other words, modes are the causes of the event, which means I can find them as an answer to the question: What caused this failure? Failures or faults may produce a few effect, but not all of them will be the cause for the Main Event. So, I am looking and analyzing the modes that have safety significance (connected with the Main Event).

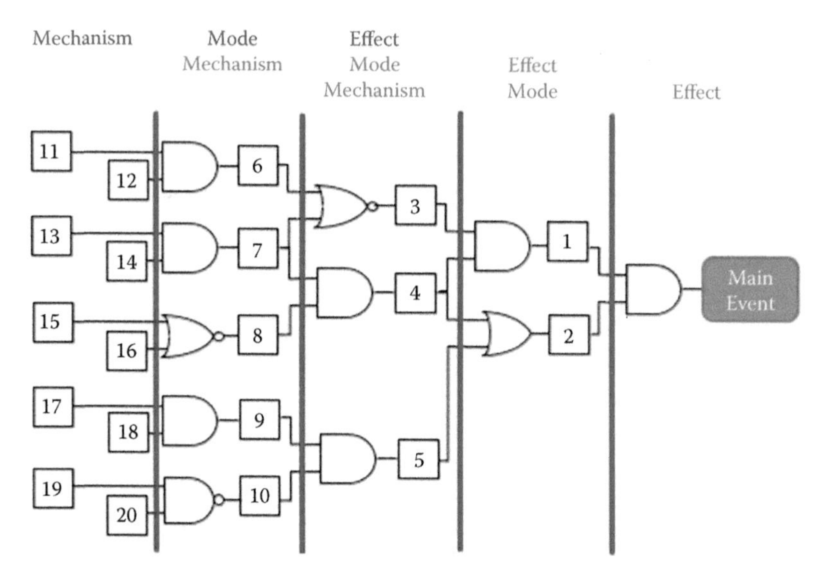

FIGURE 6.4
Connections of mechanism, mode, and effect in FTA.

Failure mechanisms are ways that show how the failure (or fault) happens, imploding into itself its cause. So, this mechanism can be found by answering the question: How it happened? Investigating the schedule (order) of events that caused Main Events, I am looking for the causes of why it happened.

Figure 6.4 shows how there is an interconnection between the mechanisms, modes, and effects in the FTA. It is important to mention here that these three terms are connected. The mechanism of the faulty equipment produces the mode, which produces the effect of failures. In general, the mechanisms are left, the modes are in the middle, and the effects are placed right in the FTA diagram.

Let us give some examples for that.

The elements of fire are material that can burn (***matter***), ***temperature***, and ***oxygen***. Fire is a chemical reaction between matter and oxygen, which is produced by high temperature. So, the ***mechanism*** of the fire is the aggregation of matter, temperature, and oxygen. The way of achieving this aggregation is the ***mode*** and the flame (the result of the fire) is the ***effect***.

If the pipe from the reservoir of my car is faulty (cut or clog), the fuel will not enter the car's carburetor, which means that the carburetor will not mix air with gasoline and it will not deliver the right mixture into the engine, meaning that the engine will not work! It means that the cutting or clogging of the gasoline pipe is the failure mechanism that stops the gasoline from entering the carburetor and mixing with the air, so no mixture of gasoline and air is produced (failure mode!) and the engine does not work (failure effect!).

Dealing with failure mechanisms, failure modes, and failure effects is important for the elimination and mitigation of the effects before the Main Event happens. When dealing with them during the FTA, we are actually decreasing the probability of the Main Event happening and, in addition, decreasing the consequences of this same Main Event, which will help us with the ETA.

7

Fault Tree Analysis

7.1 Introduction

Fault Tree Analysis (FTA) is a method that is used not only in science but also in manufacturing industry and many other areas (aerospace, nuclear, medicine, chemical, pharmacy, criminology, etc.). It is a method that was developed in 1962 by H. A. Watson from Bell Laboratories (USA) with the purpose of evaluating the Launch and Control System of Minuteman I missiles. Soon, the benefits of the FTA were recognized by Boeing (one of the companies that worked on the Minuteman missiles) and they started using it for their design and production process for their aircrafts. In the next 10 years, FTA spread in many areas, and it is also extensively used today.

The FTA investigates potential faults in complex systems by taking into consideration their mechanisms, modes, and effects and is usually used to quantify their contribution to the overall system failure (Main Event). It is best to start with the FTA during the design process because if the FTA shows that something is unreliable during design, it can be fixed without much effort and cost.

Speaking from the point of view of Logic,* the FTA is a deductive approach of analyzing the systems operations. It goes from the general to the particular or as it is well known, as the "top–down" approach. It is a scientific approach when you are proposing a theory and looking for the data that will prove that your theory is correct.

And here we have a problem.

If you do not find the data that will prove your theory is correct and if the data that are available are not enough to show you that your theory is wrong, then there is a problem. Let me explain this a little bit simply:

The deductive approach is used by detectives to solve a crime.† Let us say they are starting from a particular event (murder) when there is a dead body

* Logic in the context of this book is defined as the science that investigates the methods of thinking and reasoning based on the available data with particular quantity and particular quality (true or false).

† Of course, those detectives do not build the FTA diagram! They just use the deductive process to analyze data with the intention of finding suspects for the crimes and to produce evidence that will be used during trials.

of a person. The purpose is to find the murderer, so the detectives will try to conduct an investigation (analysis) starting with gathering as much data as possible in connection with the person murdered, such as his human and social environment, overall situation regarding the time, place, and conditions around the place where the murder happened, witnesses, and so on. Generally, there are plenty of other things that detectives want to know or find to help them find the murderer. In general, it is an iterative process: Sometimes the gathered (available) data are not enough for a reliable decision, so detectives look for more data in connection with the previously gathered information. Data are used to create a particular scenario that will determine the murderer, the motive for the crime, and how it was committed. Sometimes, all the gathered data are not enough for a reliable decision, so they cannot point at the murderer. However, this does not mean that there is no murder, because the dead body still exists.

In reality, there are a lot of scientific theories that cannot be proven, but this does not mean that they are wrong. It simply means that future development of science will (maybe) provide more data that can be taken into consideration and help to make a decision about the validity of those theories.

So, in general, FTA will not always bring us to the reliable decision-making process (is our theory OK or not OK?). The reason for that is that the process of dealing with the FTA is strongly dependent on the following factors:

- The type and characteristics of the event (theory, system, process, operation, event, etc.) which is investigated.
- The method of data gathering and the preservation of data.
- The quantity and quality of data gathered.
- The methods and tools available and used for analysis.
- The competence (knowledge and experience) of the people who are analyzing the data and those who are responsible for the decision-making.

Maybe the previous points will not present you with a warning, but do not forget that FTA is usually associated with complex systems, so the data gathering (and its analysis) is never an easy job.

7.2 Symbols Used for FT

The main symbols used for the FT construction are presented in Table 4.2. These are symbols for logical operations between the elements of the system (equipment, materials, activities, processes, etc.). They have inputs and outputs and the outputs are dependent on the states of the inputs. Whatever the input (equipment, materials, activities, processes, event, etc.), the outputs can change the overall nature of the inputs. As an example, a motor changes electrical energy into mechanical energy and a generator changes mechanical energy into electrical energy.

In addition, the symbols for the flow charts are also used in the FT. Even though the flow charts symbols are around 30, just a few of them are used in building the FT. The symbols that I choose to use are presented in Table 7.1.

There is no limitation on which symbols you may use from flow charts, but I found those in Table 7.1 to be most useful. The reason is that usually whatever you need to explain about the event (equipment, process, etc.) you may do it by writing it with the Event symbols.

The construction of the FT is staring from the Main Event. Do not forget: The FTA is a top-down approach!

I am putting Main Event and this is our failure effect that I would like to investigate. The next step is to look for modes: What can cause this effect to happen?

TABLE 7.1

Flow Charts Symbols Used in FT Constructions

Symbol	Name	Meaning
	Main event	It is symbol that is used to present Main Event in FT. In the literature, you can find it as Terminator (Start or End of flow chart).
	Event	It is used to describe events in the FT. Usually it comes after the gate symbols explaining the output of the gate operation. In flow chart, it is process or activity.
	Decision	It is used whenever the input that I need to establish in the gate depends on decision which operator (or manager) needs to do.
	Inhibit	Inhibit gate is used to present a scenario in which the output event occurs if all input events occur and an additional conditional event (usually CLK or ENABLE in digital circuits) also occurs. An inhibit gate is same as AND gate with an additional input.
3 3	**Connector**	Considering that FT is space consuming, sometimes I need to "transfer" the events on another page. This is useful especially when system is complex. Left connector is previous page and right connector is next page. So, putting the number inside the Connector, I reference the point (in our case 3) on another page where the FT diagram continues.

Note: These symbols are my choice! There is no reason not to use different symbols, but be very clear about what you are using and be sure that everyone around you understands them.

7.3 Executing FTA

Figure 7.1 shows a diagram of a general flow chart.

Red arrows are connected with the activity that causes a change in the system and green arrows are connected with the activity that does not change anything. Systems are explained in Chapter 6, so in the next paragraphs, I will speak about the steps of conducting the FTA.

7.3.1 Gathering Knowledge

I need to gather good knowledge of the system that is under consideration, and there is a good reason for that. Like a doctor who must have good knowledge of the normal functioning of a human body to provide us with the best therapy, I need to have a good knowledge of the normal functioning of a system to provide the best protection or to fix the problems when it is functioning abnormally.

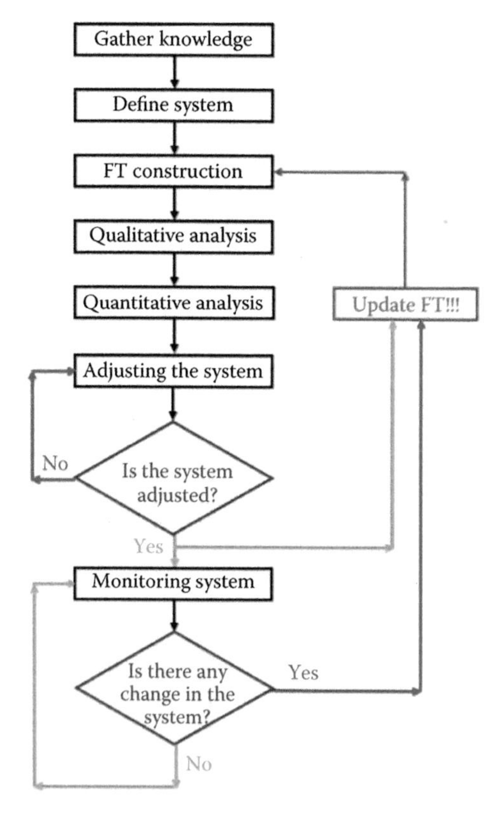

FIGURE 7.1
Flow chart for executing FTA.

In the previous paragraph, I said that the FTA is most useful when it is done during the design process. If I am implementing the FTA during the design process, by the nature of activities, I am urged to gather knowledge of all aspects of the functioning of the system under design. Speaking about all aspects, I need to take care of the purpose of the system, basic and additional functions needed to provide that purpose, components of the system (subsystems), internal relations and correlations of the subsystems, and relations and correlations with the humans and the environment where the system will operate.

But I would like to present the FTA as it is used in the functioning of a company (organization) in reality. This is an area where the FTA is used as a part of the BM for identifying hazards and risk associated with everyday operations. To gather such knowledge about the system, I need to rely on:

1. General theoretical knowledge for the methods used in the operating of the system.
2. The operating manuals for the equipment (from the manufacturer).
3. Reviews or comments from other users of the system.
4. The experience of the employees who are working with this equipment (operation, activity, process, etc.).

It is important that when you gather knowledge, you follow the same order as the one given earlier. The reason for that is that these four paths of gathering knowledge are arranged by their importance (from general to specific).

The general theoretical knowledge gives you a bigger picture of the situation. It helps you understand the general theory behind the problem and how it applies to your particular environment. It simply helps you understand why this particular solution is implemented in your case.

Operating manuals for the equipment (process) are a continuation of the general theory, the only difference being that they are more specific. They are basically the place where the theory is put into practice. Anyway, they do contain all of the required information about your equipment (process), but they do not contain all of the specific information about the behavior of the equipment (process) in correlation to your site.

Reviews or comments from other users are always useful, especially when you first start working on the system. They provide us with the data that will tell us what to expect: good or bad. This information can be gathered from Internet forums, magazines, or review websites or by establishing a direct contact with other users of the equipment.

Relying on the experience of your employees is the most important thing! My father used to say that you need to treat any piece of equipment as a woman. If you take care of it and treat it with respect, it will serve you at its best. It may sound strange, but my father, as a long time experienced engineer in industry, had an opinion that equipment (or processes) have souls. This does not mean that they have human souls, but a particular individual "behavior" that is strongly connected to the environment where the equipment (or processes) are operating as well as with the behavior of operators using the equipment and conducting the process. And this particular "behavior" can be noticed and registered only by the employees who are operating the equipment (or the process). I am very grateful to the people who have operated some equipment (or were familiar with it) because the conversations with them have helped me a lot during my engineering career. To be more precise, the conversations with them helped me understand the individual "behavior" of the equipment and to fix the emerging problems. Never underestimate the conversations with the other employees! Their help in providing valuable information is priceless!

Do not forget: The important point during the gathering of knowledge is not just to gather information, but to use this information to understand how the equipment (or process) behaves or will behave.

7.3.2 Defining the System

Here you need to define your system under consideration. All data and advices on how to define general system are given in Chapter 6 of this book, so you may use the information there to define your system. The main reminder is: Do not forget that the system under analysis is connected to your perception of how the system works (should be working), which means you are the one who is producing the model of the system. The main recommendation is to try and be as much detailed and accurate in determining the boundaries and functional blocks of the system as possible!

7.3.3 Fault Tree Construction

At the beginning, be clear what the purpose of your FT is and put the external and internal boundaries of the system (which you will analyze) on paper. And again: Be open minded! Do not hesitate to go back and adjust the perception of your system with the intention to fit reality.

Building an FT always starts with determining the Main Event.

The second step is started by building the full FT.* It is done by anticipating the primary events that are the main reason (mode, cause, etc.) for the happening of the Main Event. It is usually few primary events or a particular combination of individual events. The type of the combination is presented with a specific logic gate (AND or OR) with the help of the flow chart symbols presented in Table 7.1.

Any of these primary events could be individual or result of a particular combination of other individual events (in this case called secondary events). So, I determine these combinations and present them with logic gates and

* The expression "full FT" is used here because later I will do Modularization and Simplifications, and different (simplified) FTs can be produced there.

flow chart symbols. After looking for tertiary events, quarterly, etc., I stop when I reach the internal limits (limits of resolution) of my system or when I cannot find any other reasons (modes, causes, etc.) for the events in the FT.

The important point here is that the full FT should be done by a brainstorming session. So, whatever reason (mode, cause, etc.) for the primary (and further order) events shows up in your thoughts, just put it in the FT. Later you will see if it makes sense or not.

Be careful with the explanation of primary, secondary (and so on) events. Be more detailed, do not use abbreviations or acronyms and be more specific. This is important because the FT also will be seen and assessed by others.

7.3.4 Qualitative Analysis

The Fault Tree (FT) construction is a step that helps us "visualize" the system under analysis. Actually, by building the FT, I build a model of my perception of the system. The interesting point is that different people will construct different FTs because everyone has different perception of the system at hand. The reason for that is that we are all humans, meaning that we often look at things from our point of view, which is actually an expression of our knowledge, experience, personality, and social status. And this is something really beautiful and valuable about humanity: We are all different!

Unfortunately, this is not a holistic approach. It is a simplified and one-sided approach and as such does not provide a real picture of the system. That is the reason why whenever you produce an FT, you need to share it with the other employees and ask for their comment, advice, or suggestion. There is nothing wrong with that and there lies a great benefit: Others can "open your eyes" to something that you have been missing. Try to choose experienced employees from different areas that are connected with the system in different ways. Every one of them will offer you a different perspective and different context. And this is actually the most important point: You need to be able to recognize different contexts and be able (by using them) to make further optimization of the FT.

Do not underestimate anything! Try to "change your point of view" when you are analyzing other comments, advices, or suggestion. If necessary, do a few FTs with different perspectives or contexts and analyze every one of them. The results will help you find the best one. Whatever others say to you, you are the one in charge with the FT and with analysis. Others will just help you to "polish the surface."

Please note that the FT depends on the nature and context of analysis. By nature, I mean that you can be building an FT to understand a certain process or system, to find the influence of something over something, to find the probability of something to happen, and so on. The context is connected with the level of accuracy and precision that I am looking for from the FT. Of course, the context of the FTs for scientific purposes will differ from the context of the FTs for manufacturing purposes. Science does not compromise with economy, but the production on the other hand, must. Also, different industries apply different contexts. Even two companies from the same industry may apply different contexts. Same thing applies for quality and safety: The FTs for quality are different from the ones for safety. So, be careful!

An FT is actually a graphical presentation of a particular Boolean formula. Once I construct the full FT, I need to check if it is okay (Is it correct?). This is part of the qualitative analysis in which all operations and connections are checked. Here I reconsider all primary, secondary, and so on events (checking their integrity). If some of them do not make sense, they can be deleted, but be careful: Do not do it until you hear the opinion of others working with you! Their opinions, advices, and comments are very important.

Here you can do a Modularization of the full FT. Modularization is actually of the simplification part of the FTA. It is an activity of grouping particular combinations of events into a new FT called Sub-Fault Tree (SFT). This is strongly recommended for complex systems that would usually produce complex FT. Dealing with SFTs instead of FTs is good because it allows us to present more details of the system, improving our picture of the operation of the system and causes of the Main Event. It is always easier to deal with an SFT than with an FT, but it must be done very carefully: There is a danger of losing the "big picture" of the Main Event! The Main Event is an equipment, operation, process, activity, event, and so on that is a result of different processes inside the system, interactions within them, and with the environment, and because of this, it needs a holistic approach, not partial.

If everything is OK with the diagram, I will then produce an expression (a Boolean function) that describes the combinations and interdependencies inside the FT.* I will construct a formula that is an empirical presentation of the graphically presented FT. Producing the formula means that I have to apply the processes for Simplification by using the Boolean laws (Table 4.3) to obtain the minimal cut set (MCS) for the Main Event.

The MCS is basically a formula of the Main Event that is built on the combination of the first-order, second-order, or higher order MCSs. This formula can be used to eliminate or mitigate hazards or risks connected with the Main Event. The main point here is that if there are first-order MCS in the system, they are single-point failures† (SPFs). It means that the happening of the Main Event is dependent on a single subsystem (component, process, or activity). The main rule in safety is: Do not allow SPF to exist because this means that a single component can trigger the Main Event! So, having such a situation, I should consider changing something in the system with the purpose of eliminating the SPF. It can be done in different ways‡ and you should use the one that is best for your particular system.

* You do not work with SFTs here! The formula should be produced only by using FTs!
† See paragraph 4.6 (Minimum Cut Set).
‡ More on that in Chapter 12 (How to improve safety with the BM).

Producing the MCS means that the previous FT will be changed with an equivalent one, presented by a simplified MCS. Now I am changing the FT in accordance with the produced MCS, but this does not mean that I need to forget the full FT. If there is a need to make some change in the system in the future, the already produced MCS and its FT will not be valid any more. This is a situation where I implement the change into a system and this change will affect my full FT. In general, all changes shall be introduced in the full FT. Of course, now it will be very easy to put the changes inside the FT and when finished, I will produce a new MCS expression for the system with the changes implemented along with a new FT as a result of the new MCS.

7.3.5 Quantitative Analysis

When the qualitative analysis shows that the FT is a truly realistic model of the system under investigation, then I may assign particular probabilities for every event in the FT. If there are SFTs, then I associate the probabilities of the events there and put them in the full FT later.

The problem here is to determine the probabilities* of the previous events. To do this, you need a better understanding of failure mechanisms, failure modes, and failure effects. An extensive knowledge of reliability is very important to properly calculate the probabilities. But, reliability is not simple!

If I am not familiar with the system under analysis, I will try to find data to help me find the probabilities of the events inside the system. Doing this is not always easy, so such a situation is not always effective. Please note that the values of probabilities that I use in this book are my own calculations made up from the data I gathered from different sources. During this process, I had to make some decisions that I thought were pragmatic. The main point here is that using the FTA method and assigning the data inside the FT are two completely different things. Someone may find the data that I am using in the calculations strange, but this does not matter for the purpose of this book. The data here are used with the intention to explain how to use data during the calculations.

The reliability calculations are a necessity and having a clear picture of the reliability of the equipment (provided by the manufacturer) is extremely helpful. In addition to the equipment reliability data, you have to build your own database for "human reliability"† too. These data should be about human errors in both your particular industry and your company. This is the main reason why I believe that the execution of the BM should not be done by hiring an external company (as a service), but by the Quality or Safety Manager inside the company.

* Severity is connected with the Main Event and I will need it later, for the ETA.

† For the time being, reliability is a characteristic of equipment. I strongly recommend using it on services offered and on humans too.

A good Quality or Safety Manager must know the behavior of the employees and equipment inside his company and should build a database for everything that happens in it. But that is not all! He should follow events in the industry as well as learn from the mistakes or good actions of others. Do not forget: Safety shall be predictive!

In general, not having knowledge gathered from "first hand" is not good. Bad things do not choose the way they happen, so you need someone familiar within the company to get a track of how things happen.

Always have in mind that building the FT is just the first step and as such, it is only the beginning in the execution of the FTA! The integrity of the FTA is not only built by an accurate FT but also by the integrity and reliability of the data used.

7.3.6 Adjusting the System

When I finish with the analysis, I use the results to see and decide how I can improve my system. How, where, and when I can do it must be part of the decision-making process. Improvement means that I can try to adjust some of the "variables" in the system with the intention of eliminating some risks and/or mitigating some internal effects. This will result in a better and safer system. Actually, adjusting means that we introduce some change in the system that will eliminate or mitigate some of the internal effects. These changes can be putting a new circuit or piece of equipment, using better equipment or better materials, introducing new procedures, making a change in organization, and so on.

The main point here is that all these changes will also change the structure of my system, which means that the already finished FTA is not the same anymore. The changes I have included might produce some new hazards (you never know!), but they will surely change the existing values of the risk probabilities in the previous FTA.

So, I have to go and construct the FT again with the newest changes in my system. When you gather enough experience with the FTA, you will not need to repeat all steps from Figure 7.1, but do not forget: You are human and mistakes can be easily made! So, follow the procedure, repeat everything. It should go fast anyway.

This step is usually forgotten, and as far as my experience goes, many of the safety and/or quality managers do not follow it.

7.3.7 Monitoring the System

When I finish the FTA, the system should be monitored, and there are two reasons for that. The first one is to be sure that the system is continuing its normal operation and to assure ourselves that the FTA was good. The second reason is that we can never really adjust the system to mitigate or eliminate all risks, so I need to monitor it all the time. In addition, the monitoring of the system will help me deal with some hidden risks that were not registered before.

To have the safest road traffic, everyone would need to drive armored vehicles (tanks). They are very slow and they shield us from any harm. But this traffic is not really useful, so we are driving cars (as they are!) and looking through the windshield all the time, following the traffic situation in real time. This helps us to adjust our driving to the present situation and undertake a particular action if necessary. This kind of monitoring must be implemented to the systems too, so we can gain information on how our system is coping with the external and internal influences.

Any notice regarding any misbehavior of the system (abnormal operation) should trigger a particular action by the operator. This action can be "acute,"* but it is better if I can produce a systematic solution for the problem registered (in case it happens again). This is not always possible, but anyway, in such a situation, a backup plan or a recovery plan must be available.

* "Acute" in this context means "one time action only for this particular case." If I use it always when this case happens, then it is going to be a "systematic" case, which means I need to produce a procedure.

8

FTA for Instrumental Landing System (ILS)

8.1 Introduction to ILS

Instrumental Landing System (ILS) is a ground-based navigational aid used in aviation. It is an electronic equipment that sends navigational data to the aircraft through radio signals. So, ILS is a transmitting system on the ground and there an appropriate receiver in the aircraft is adjusted to the frequency of the ILS signal. The signal is received by the receiver in the aircraft, processed, and shown on the cockpit display, so the pilot can use it for navigation during the landing of the aircraft on the runway. In addition, this signal can be submitted to the Flight Management System (FMS, known as auto-pilot) and can be used for automatic landing purposes. Use of ILS is not very simple because there are three categories of operations and associated equipment (Facility Performants Categories CAT I, CAT II, and CAT III), and all of them have different specifications (CAT III is the strongest by requirements).

The ILS contains three types of equipment*: A Localizer (LLZ) used to give horizontal guidance (left or right from the axis of the runway), a Glidepath (GP) used to give vertical guidance (up or down from the ideal landing angle), and three Markers (M) that provide the pilot with the information about the distance from the threshold of the runway. In addition, all these three pieces of equipment are doubled and each of them has doubled monitoring and control devices to improve the integrity and reliability of the equipment. So, the ILS provides three-dimensional navigation information to the aircraft. Considering that weather can be very hostile, the ILS is extremely useful in bad weather conditions and during the night (poor visibility situations).

The document that posts requirements for the navigational signal, so called, Signal in Space (SiS) is Annex 10 (Aeronautical Telecommunications), Volume I (Radio Navigation Aids), published by the International Civil Aviation Organization (ICAO). Other documents issued by the Radio Technical

* When I mention the ILS in this book, it means that I am talking about the characteristics of LLZ, GP, and Markers in general.

Commission for Aeronautics (RTCA) and the European Organization for Civil Aviation Equipment (EUROCAE) are dedicated to producing Acceptable Means of Compliance for the aviation equipment as Minimum Operating Performance Standards (MOPS) and Minimum Avionics System Performance Standards (MASPS).

I will not go into details about the ILS, but I will use one of the three parts of ILS (LLZ) to present the FT of Main Event called "No SiS."* Not having SiS means that the navigation receiver in the aircraft is not receiving a good navigation signal or not receiving an ILS signal at all. If the signal present is not good, it means that the availability, reliability, integrity, and continuity of the service of the operation are not good either. In such a condition, in bad weather conditions and during low visibility (fog, night, rain, etc.), the pilot is not able to use any navigation aid to help him or her land on the runway. This means that it is extremely dangerous to land on the runway because the pilot cannot see the real position of the aircraft in regard to the runway position.

8.2 Determining the System and Its Boundaries

I will produce a Fault Tree (FT) only for the Localizer (LLZ) for CAT III Facility Performance equipment.

The internal boundaries will be transmitters, monitors, antenna relay, and antenna. In this case, it is better to choose internal boundaries that hold us from going too much into detail. Later, if needed, I can move the internal boundaries to the Line Replaceable Units (LRUs) of the transmitters and monitors by producing Sub-Fault Trees for them. It is reasonable to do that because when I am buying the ILS, the data for its reliability and other specifications for the equipment can be obtained from the manufacturer. Anyway, these data (for your equipment) must be proved on your site under your operational environment, so you need to have a procedure to monitor and record the data necessary to prove these specifications. This is a regulatory requirement for aviation equipment.

The external boundaries will be from the building (housing, shelter) where the LLZ is installed to the outside area where the LLZ antenna (20–30 m from the LLZ building) is installed together with the field monitoring antenna (150 m in front of the LLZ antenna).

* "No SiS" has a broader definition than "signal is not being radiated." Actually, "No SiS" also means that the signal that is received by the aircraft navigational receiver cannot be used to provide navigation (something is wrong with it!).

The important point to mention here is that the whole ILS is completely automatic, so human involvement is limited to only monitoring and maintenance. Regular maintenance is done by following the manufacturer's procedures, so the influence of humans on the equipment is very low.

8.3 Gathering Knowledge for LLZ

The block diagram of a Localizer is given in Figure 8.1.

The LLZ and GP are similar in their functioning: They both have two transmitters, one system of transmitting antennae connected to the transmitters through a relay and distribution box, two monitors, and one field monitoring antenna.* The relay is controlled by monitors and used to replace the faulty transmitter with a normal one, connecting it to the antenna system. The Distribution box is distributing the transmitting signal to the antenna system, built by individual antennae whose number varies from 12 to 24 for LLZ (I will use 12) and from 2 to 3 for GP. The difference is that the LLZ antennae provide horizontal guidance, so they are mounted horizontally, and the GP antennae provide vertical guidance, so they are mounted vertically on the mast.

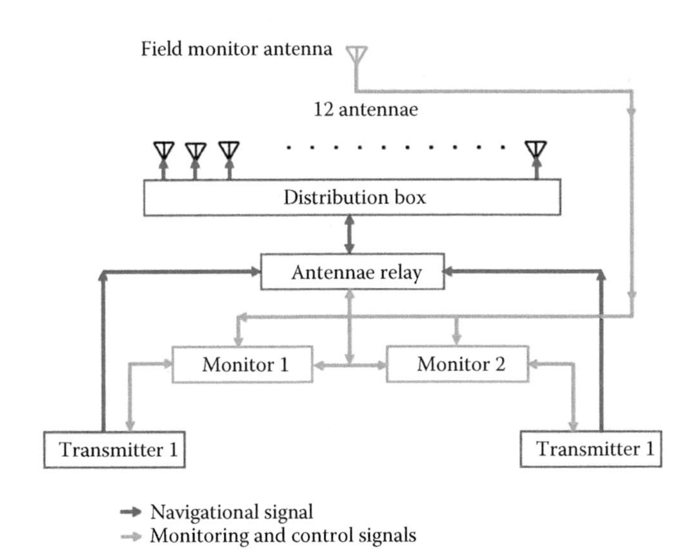

FIGURE 8.1
Block diagram of Localizer (LLZ).

* There is possibility for another monitor, so-called Far Field Monitor (FFM), but it is optional!

Both monitors monitor the functioning of both transmitters and every monitor monitors itself and the other monitor. The overall monitoring and control is supported by software.* So, in the case of any type of malfunctioning, they are able to react:

- If one transmitter is faulty, the monitors will notice that and they will switch off the faulty one and connect the second (normal) transmitter to the antenna.
- If the other transmitter is also faulty, then they will switch the ILS off.
- If something is wrong with one of the monitors, there is an algorithm to determine the faulty one and to switch it off.
- If both monitors are faulty, the ILS will be automatically switched off.
- If the field monitoring antenna cannot detect any SiS, then the monitors will switch off the ILS.

Information for all these events will be transferred to the engineering control room and to the Air Traffic Controller (ATCo) on duty, so everyone (pilot, ATCo, and maintenance engineer) will be informed on time. Rules require that the time from the fault of equipment to the switch off of the transmitters should not be more than 1 s for the CAT III LLZ. This time is enough for pilot to react.

8.4 Fault Tree for LLZ CAT III

The Fault Tree for the Main Event is called "No SiS" for CAT III LLZ and given in Figure 8.2.

Looking at the diagram, you can notice that there are a few causes for the No SiS for the LLZ. These are:

1. Transmitter 1 (T1) and Transmitter 2 (T2) are not working. Monitors will register this and they will switch off the LLZ because both transmitters are out of service (faulty!).
2. Transmitter 1 (T1) is not working and Transmitter 2 (T2) is working, but Monitor 1 (M1) and Monitor 2 (M2) are not working. When both monitors are not working, they cannot monitor the transmitters. This means that the integrity of the SiS is endangered, so the LLZ will be switched off.
3. Transmitter 1 (T1) is working and Transmitter 2 (T2) is not working, but Monitor 1 (M1) and Monitor 2 (M2) are not working. When

* Please note that software will not be considered here because it is embedded in the equipment.

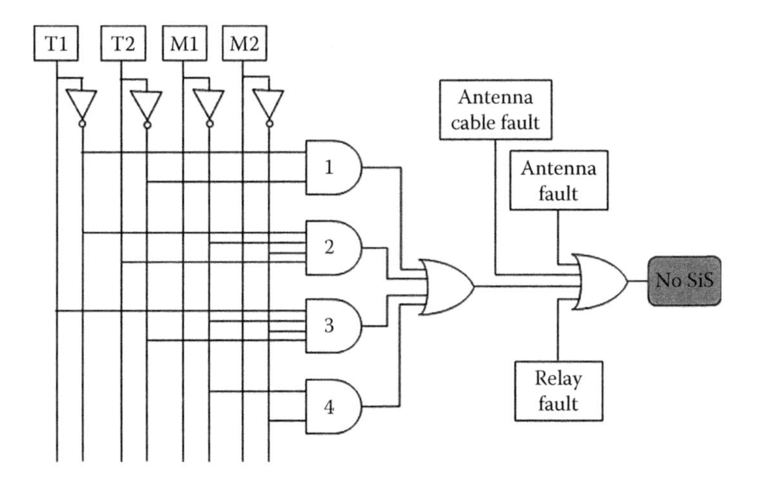

FIGURE 8.2
FT for LLZ CAT III (No signal in space—No SiS).

both monitors are not working, then the integrity of the SiS is endangered, so the LLZ will be switched off.

4. Monitor 1 (M1) and Monitor 2 (M2) are not working. When monitors are not working, the integrity of the SiS is endangered, so the LLZ will be switched off.
5. The antenna is faulty, both the transmitters and monitors are functioning normally, but the signal is not radiated because of the faulty antenna.
6. The antenna cable is faulty. Both the transmitters and monitors are functioning normally, but the signal is not radiated because of the faulty cable (the signal does not reach the antenna!).
7. The antenna relay is faulty. Both transmitters and monitors are functioning normally, but the signal is not radiated because of the faulty relay that does not connect the transmitter to the antenna through the antenna cable.

Please note that the Distribution box consists of cables that distribute the signal from the transmitter to the different antennae in the antenna system. So, the fault of Distribution box is covered by the fault of the antenna cable. Also, consider that if one antenna (the system is built from 12 antennae) is faulty, then the whole antenna system is faulty.

As you can notice, all of the situations with the transmitters and monitors are presented with AND gates (1–4) because they need to happen at the same time. All other connections are with OR gates because it is enough if only one of them happens.

8.5 Qualitative Analysis

Qualitative analysis means that I need to check the overall situation again in order to check the correctness and accuracy of my diagram. I do not have any reason not to believe that everything is OK because I already checked it a few times, so I will continue with the analysis.

The next step is to produce the Boolean function as a sum of minterms from the diagram in Figure 7.2. It will be:

$$F = \overline{T1} \bullet \overline{T2} + \overline{T1} \bullet T2 \bullet \overline{M1} \bullet \overline{M2} + T1 \bullet \overline{T2} \bullet \overline{M1} \bullet \overline{M2} + \overline{M1} \bullet \overline{M2} + \overline{AR} + \overline{AC} + \overline{AN}$$

The antenna relay, antenna cable, and antennae are independent from the transmitters and monitors, so I can just not take them into consideration for now. I will try to use the rules in Table 4.3 to simplify the following Boolean function:

$$F = \overline{T1} \bullet \overline{T2} + \overline{T1} \bullet T2 \bullet \overline{M1} \bullet \overline{M2} + T1 \bullet \overline{T2} \bullet \overline{M1} \bullet \overline{M2} + \overline{M1} \bullet \overline{M2}$$

Actually, I am trying to produce the Minimal Cut Set (MCS) for the Boolean function F (No SiS) through simplification.

The first step is to group the minterms:

$$F = \left[\overline{T1} \bullet \overline{T2} + \overline{T1} \bullet T2 \bullet \overline{M1} \bullet \overline{M2}\right] + \left[T1 \bullet \overline{T2} \bullet \overline{M1} \bullet \overline{M2} + \overline{M1} \bullet \overline{M2}\right]$$

Then, I take out the same elements from the two brackets and put them in the front (extracting the common multiplier in front of the bracket):

$$F = \overline{T1} \bullet \left[\overline{T2} + T2 \bullet \overline{M1} \bullet \overline{M2}\right] + \overline{M1} \bullet \overline{M2} \bullet \left[T1 \bullet \overline{T2} + 1\right]$$

Now, for the first bracket, I apply the ***Combination of operations*** (the upper part of Table 4.3) and for the second bracket, the 1 will be transformed by ***Complementation*** (the right part of Table 4.3). The result is:

$$F = \overline{T1} \bullet \left[\overline{T2} + \overline{M1} \bullet \overline{M2}\right] + \overline{M1} \bullet \overline{M2} \bullet \left[T1 \bullet \overline{T2} + T2 + \overline{T2}\right]$$

The next step is to apply the ***Combination of operations*** (the upper part from Table 4.3) for the second bracket:

$$F = \overline{T1} \bullet \left[\overline{T2} + \overline{M1} \bullet \overline{M2}\right] + \overline{M1} \bullet \overline{M2} \bullet \left[T1 + T2 + \overline{T2}\right]$$

Implementing the **Idempotent law** (second part) from Table 4.3, I get:

$$F = \overline{T1} \bullet \left[\overline{T2} + \overline{M1} \bullet \overline{M2}\right] + \overline{M1} \bullet \overline{M2} \bullet [T1 + 1]$$

Multiplying the elements outside the brackets with those in brackets, I get:

$$F = \overline{T1} \bullet \overline{T2} + \overline{T1} \bullet \overline{M1} \bullet \overline{M2} + T1 \bullet \overline{M1} \bullet \overline{M2} + \overline{M1} \bullet \overline{M2}$$

Let us group the two elements in the middle again:

$$F = \overline{T1} \bullet \overline{T2} + \overline{M1} \bullet \overline{M2} \bullet \left(\overline{T1} + T1\right) + \overline{M1} \bullet \overline{M2}$$

The sum in the bracket is 1 (***Complementation***), so:

$$F = \overline{T1} \bullet \overline{T2} + \overline{M1} \bullet \overline{M2} + \overline{M1} \bullet \overline{M2}$$

Applying the ***Idempotent law*** from Table 4.3 one more time, I get the final Boolean formula or the MCS:

$$F = \overline{T1} \bullet \overline{T2} + \overline{M1} \bullet \overline{M2}$$

The calculated MCS is an extremely logical result: No SiS can happen only if both the transmitters or both the monitors are faulty at the same time (under the assumption that nothing is wrong with antenna, antenna cable, and antenna relay). So, if I would like to be sure that the SiS is being transmitted to the aircraft, I will need at least one transmitter and one monitor to work!

It is easy to show that this is the MCS and for that purpose I use the sum of minterms for function F and produce a table with all possible combinations (Table 8.1) for the function calculated from Figure 8.2.

TABLE 8.1

Finding the Sum of Minterms for $F = \overline{T1} \bullet \overline{T2} + \overline{M1} \bullet \overline{M2}$

M_2	$\overline{M_2}$	M_1	$\overline{M_1}$	T_2	$\overline{T_2}$	T_1	$\overline{T_1}$	$T_1 \bullet \overline{T_2}$	$M_1 \bullet M_2$	$T_1 \bullet T_2 + M_1 \bullet M_2$	Minterms
0	1	0	1	0	1	0	1	1	1	1	$\overline{T_1} \bullet \overline{T_2} \bullet \overline{M_1'} \bullet \overline{M_2}$
0	1	0	1	0	1	1	0	0	1	1	$T_1 \bullet \overline{T_2} \bullet \overline{M_1'} \bullet \overline{M_2}$
0	1	0	1	1	0	0	1	0	1	1	$\overline{T_1} \bullet T_2 \bullet \overline{M_1'} \bullet \overline{M_2}$
0	1	0	1	1	0	1	0	0	1	1	$T_1 \bullet T_2 \bullet \overline{M_1'} \bullet \overline{M_2}$
0	1	1	0	0	1	0	1	1	0	1	$\overline{T_1} \bullet \overline{T_2} \bullet M_1 \bullet \overline{M_2}$
0	1	1	0	0	1	1	0	0	0	0	
0	1	1	0	1	0	0	1	0	0	0	
0	1	1	0	1	0	1	0	0	0	0	
1	0	0	1	0	1	0	1	1	0	1	$\overline{T_1} \bullet \overline{T_2} \bullet \overline{M_1'} \bullet M_2$
1	0	0	1	0	1	1	0	0	0	0	
1	0	0	1	1	0	0	1	0	0	0	
1	0	0	1	1	0	1	0	0	0	0	
1	0	1	0	0	1	0	1	1	0	1	$\overline{T_1} \bullet \overline{T_2} \bullet M_1 \bullet M_2$
1	0	1	0	0	1	1	0	0	0	0	
1	0	1	0	1	0	0	1	0	0	0	
1	0	1	0	1	0	1	0	0	0	0	

Looking at the table, I construct the following sum of minterms:

$$F = \underbrace{\overline{T_1} \bullet \overline{T_2} \bullet \overline{M_1} \bullet \overline{M_2}}_{1} + \underbrace{T_1 \bullet \overline{T_2} \bullet \overline{M_1} \bullet \overline{M_2}}_{2} + \underbrace{\overline{T_1} \bullet T_2 \bullet \overline{M_1} \bullet \overline{M_2}}_{3} + \underbrace{T_1 \bullet T_2 \bullet \overline{M_1} \bullet \overline{M_2}}_{4} + \\ + \underbrace{\overline{T_1} \bullet \overline{T_2} \bullet M_1 \bullet \overline{M_2}}_{5} + \underbrace{\overline{T_1} \bullet \overline{T_2} \bullet \overline{M_1} \bullet M_2}_{6} + \underbrace{\overline{T_1} \bullet \overline{T_2} \bullet M_1 \bullet M_2}_{7}$$

As you can notice, there are seven minterms that produce a sum of minterms. I try to simplify them and the first step is to group 1 and 2, 3 and 4, and 5 and 7. The result is:

$$F = \underbrace{\left(\overline{T_1} + T_1\right) \bullet \overline{T_2} \bullet \overline{M_1} \bullet \overline{M_2}}_{1+2} + \underbrace{\left(\overline{T_1} + T_1\right) \bullet T_2 \bullet \overline{M_1} \bullet \overline{M_2}}_{3+4} + \underbrace{\left(\overline{M_2} + M_2\right) \bullet \overline{T_1} \bullet \overline{T_2} \bullet M_1}_{5+7} \\ + \overline{T_1} \bullet \overline{T_2} \bullet \overline{M_1} \bullet M_2$$

Applying **Complementation** (from Table 4.3), I get:

$$F = \overline{T_2} \bullet \overline{M_1} \bullet \overline{M_2} + T_2 \bullet \overline{M_1} \bullet \overline{M_2} + \overline{T_1} \bullet \overline{T_2} \bullet M_2 + \overline{T_1} \bullet \overline{T_2} \bullet \overline{M_2}$$

By doing the grouping (first + second and third + fourth), I get:

$$F = \left(\overline{T_2} + T_2\right) \bullet \overline{M_1} \bullet \overline{M_2} + \overline{T_1} \bullet \overline{T_2} \bullet \left(M_2 + \overline{M_2}\right)$$

Using *Complementation* again, the final result is the same as the previously calculated one:

$$F = \overline{M_1} \bullet \overline{M_2} + \overline{T_1} \bullet \overline{T_2}$$

To get full MCS, I add the antenna relay, antenna cable, and antenna, so the overall qualitative analysis results in the following formula:

$$F = \overline{T1} \bullet \overline{T2} + \overline{M1} \bullet \overline{M2} + \overline{AR} + \overline{AC} + \overline{AN}$$

I notice that this presentation of MCS (No SiS) consists of a sum of five minterms: two of second order and three of first order. As mentioned in paragraph 4.6 (Minimal Cut Set), the first-order MCS elements are known as single point failures (SPF) and are critical for the functioning of every system. This is because if one of them fails, the whole system fails. So, in this case, I have a Localizer worth approximately 500,000 USD, but one antenna cable or antenna relay (worth maybe less than a hundred USD) can endanger the functioning of the whole ILS.

8.6 Quantitative Analysis

8.6.1 Characteristics for SiS for LLZ CAT III

When I have the MCS, I try to calculate the overall probability of having No SiS.

But the situation with the ILS is not simple at all. There are four operational specifications for the ILS that must be satisfied to provide the required safety of landing operations. These are Availability, Reliability, Integrity, and Continuity of Service, and all of them are defined in the ICAO Annex 10, Volume 1 document.*

Availability is defined as "a percentage, may be expressed in terms of the ratio of the actual operating time divided by the specified operating time taken over a long period." Or it can be expressed by the formula:

$$\text{Availability} = \frac{\text{AOT}}{\text{SOT}} \bullet 100\%$$

where:

AOT is the Actual Operating Time
SOT is the Specified Operating Time

AOT is the time when the ILS is providing a SiS, and it is less than the SOT due to failures and time reserved for regular and nonregular (incidental) maintenance.

The Reliability for ILS is defined† as "probability that the facility will be operative within the specified tolerances for a time t, also referred to as probability of survival, P_s."

The Integrity of ILS signal is defined as "That quality which relates to the trust which can be placed in the correctness of the information supplied by the facility. The level of integrity of the localizer or the glide path is expressed in terms of the probability of not radiating false guidance signals." In other words, the integrity of the equipment is the same as that of a person: My trust in him or her is based on his or her previous actions and history of his or her deeds!

The Continuity of Service (CoS) for the ILS is expressed as "That quality which relates to the rarity of radiated signal interruptions. The level of continuity of service of the localizer or the glide path is expressed in terms of the probability of not losing the radiated guidance signals."

* Annex 10 (Aeronautical Telecommunications) to the Convention of International Civil Aviation Volume 1 (Radio Navigation Aids), 6th Edition, July, 2006, published by International Civil Aviation Organization, Montreal, Canada.

† There are no differences in the definition of reliability from the one in Chapter 5.

There are few other points that must be considered with regard to these four operational specifications.

The first point is that all of them are connected and close to each other. Different navigational aids have different values for all of them.

The second point is that for us, reliability is actually very useful. But I need to be careful how I calculate it. Do not forget that reliability is the probability of something for a particular period of time and this period of time should be carefully determined while calculating this probability knowing the MTBF.

Actually, the simplest by definition and most important for flight operation is Availability. Having an aircraft on landing and an ILS that is not working means that the Availability is 0 and because of that, all three other specifications must also be 0. This is the worst situation!

8.6.2 Probabilities for Transmitters and Monitors*

As it is written in ICAO Annex 10, Volume 1†:

> The probability of *not radiating false guidance signals* shall not be less than $1–0.5 \times 10^{-9}$ in any one landing for Facility Performance Categories II and III localizers.

In addition, there is another rule:

> The probability of *not losing the radiated guidance signal* shall be greater than $1–2 \times 10^{-6}$ in any period of 30 s for Facility Performance Category III localizers intended to be used for the full range of Category III operations (equivalent to 4,000 h mean time between outages).

Please note that the first probability is for "not radiating" and the second one is for "not losing" (if it is already radiated!). So, the first probability deals with the fault of monitors (which do not register the radiation of the false signal!) and the second probability deals with the fault of transmitters (they were radiating, but they suddenly stopped!). Also, the term "mean time between outages" (MTBO) is used, which means that the signal is connected with outages and the equipment is connected with failures. In general, there are a plenty of "variations" used by companies to deal with the reliability of equipment and services. For a single transmitter MTBO = MTBF, but for a double configuration of the transmitters MTBO > MTBF. Anyway, different subjects have different definitions for the MTBO.

I will use both values given in the ICAO Annex 10 Volume 1 for our Main Event "No SiS" from the LLZ. So, the probability that our monitor will detect a false signal must be less than 0.5×10^{-9}. In other words, the

* How I can use the reliability for the probability of faults is explained in Chapter 5.

† For a more detailed information, see paragraph 3.1.3.12 in this document!:

probability of radiating good SiS from the LLZ transmitters should be more than 0.9999999995.* The probability of a transmitter fault must be less than 0.2×10^{-5}. In other words, the probability of the transmitters working properly must be 0.999998.

These are regulatory requirements, but I need to calculate the probability of my transmitters and monitors. For this purpose, I will use the formula for reliability that can be found in the beginning of paragraph 5.2.

Using this formula is okay, but there is problem when choosing the time t. In the formula, t is the time for which I would like to calculate the reliability of the equipment. The ILS is working nonstop. There is no rule on switching it on or off, which means that the ILS is working 24/7. The reason for that is that this is the optimal solution for a stable operation of the equipment. But, the pilots are only using ILS (LLZ) during the landing of the aircraft. The landing speed of aircraft varies from 210 to 290 km/h, depending on the size of aircraft (heavier aircraft need more speed during landing). Taking into account the lower speed (the aircraft will need the SiS for a longer time!) and that the coverage of the ILS is 25 NM (1 NM = 1.852 km, so 25 NM = 46.3 km), I can calculate the time of landing in the worst-case scenario for a low-speed aircraft:

$$t = \frac{S}{v} = \frac{46.3\ \text{km}}{210\ \text{km/h}} = 0.22\ \text{h}$$

where:

S is the distance of using the ILS before landing
v is speed of the aircraft (worst-case scenario)

Actually, 0.22 h is actually 13.2 min, but do not forget that MTBF is expressed in hours, so t must also be expressed in hours.

Using these values for t and the MTBF of the LLZ, I can calculate the probability of the LLZ not providing SiS during the landing using the formula for $P_{(faulty)}$ from earlier.

$$P_{(faulty)} = 1 - e^{-\frac{t}{MTBF}} = 1 - e^{-\frac{0.22}{4000}} = 1 - 0.99994496676 = 5.5 \bullet 10^{-5}$$

But this calculation uses an MTBF of 4,000 h. This requirement is mentioned in a few aviation regulatory documents for a 2-year initial certification of the equipment.

Reality is different, but different in a good way! I contacted a few companies (manufacturers of ground-based navigational equipment) regarding the

* Having "No SiS" does not mean that aircraft automatically will have an accident. There are other activities that influence the aircraft in such a situation, and some of them are related to the pilot (will he notice this, in which stage of the flight will it happen, what actions will he undertake, etc.), but the aircraft accident has bigger probability to happen (3×10^{-9}).

MTBF of their equipment, asking them for the MTBF of their ILSs and some of them responded to my question. From the data that I got, the value of the MTBF of LLZ extends from 26,000 to 33,000 h.* Considering that a bigger MTBF means better equipment, I choose worst case of MTBF of 26,000 h.

Using this value, I get a probability equal to:

$$P_{(faulty)} = 1 - e^{-\frac{t}{MTBF}} = 1 - e^{-\frac{0.22}{26000}} = 1 - 0.99999153551 = 8.46 \bullet 10^{-6}$$

I use this value as the probability of failure of transmitters and monitors of the LLZ because I do not have the individual values for the calculation. So, it is:

$$\overline{T1} \bullet \overline{T2} + \overline{M1} \bullet \overline{M2} = 8.46 \bullet 10^{-6}$$

8.6.3 Other Probabilities

The three other probabilities included into the formula are the probabilities of fault of antenna, antenna cable (with connectors), and antenna relay.

The LLZ antenna is actually a system of 12 antennae that are simple dipoles or logarithmic antennae and mounted in one horizontal line on masts at a constant height. The signal in the space is a combination of all the signals transmitted by different antennae and it is distributed by a particular distribution box that can also be a part of the transmitter. The system does not have moving parts and it is usually made of aluminum. It is strong, protected by plastic, and extremely durable. If it is used in accordance with the specifications, it cannot be damaged and is not prone to wearing because there are no moving parts. So, I can assume that the probability of fault of the antennae is 0.

The antenna cable is usually a coaxial cable due to the frequency used for the LLZ (VHF band). Together with the connectors, they are critical for the reliability of the system. They are sensitive to mechanical stresses (temperature changes, air humidity, air pressure, shocks, and vibrations) and electromagnetic stresses caused by electromagnetic environment. Due to these sensitivities, it can fail, especially in places where the connectors are connected to cables. The best cables and connectors available are used (military specification applies here!) for the ILS (and aviation in general!). Considering that these cables are fixed to something (no movement), the faults that could potentially happen at places where the connectors are connected to the cables can be prevented by having a good protection against humidity (during installation) and regular maintenance (during operation).

* For real cases (equipment installed on aerodrome), I must know the MTBF of every part! For the purpose of this book, I choose the worst-case scenario, which is the lowest MTBF.

I contacted a few cable manufacturers and all of them told me that if the cables and connectors are used in accordance with the specifications, they cannot fail. But, they mentioned that the MTBF is between 50,000 and 100,000 h. For my calculations, I use the MTBF of 50,000 h, as it is the most critical.

$$P_{(faulty)} = 1 - e^{-\frac{t}{MTBF}} = 1 - e^{-\frac{0.22}{50000}} = 1 - 0.9999956 = 4.4 \bullet 10^{-6}$$

The antenna relay is a solid-state UHF relay that is quite different from the relays for DC and low-frequency AC. Using the US Department of Defense document MIL HDBK 217F (Reliability Prediction for Electronic Equipment), I calculated that the value of 1.2 faults in 10^6 hours (MTBF) is equal to:

$$MTBF = \frac{AOT}{n} = \frac{1000000}{1.2} = 833333\ h$$

Using the formula for $P_{(faulty)}$, t = 0.22 h, and MTBF = 833333, I see that the probability for the relay to be faulty is equal to:

$$P_{(faulty)} = 1 - e^{-\frac{t}{MTBF}} = 1 - e^{-\frac{0.22}{833333}} = 1 - 0.999999736 = 2.65 \bullet 10^{-7}$$

8.6.4 Final Calculation

The final calculation should gather all values, so the formula for F is:

$$F = \left(\overline{T1} \bullet \overline{T2} + \overline{M1} \bullet \overline{M2}\right) + \overline{AR} + \overline{AC} + \overline{AN}$$

Now, I insert the calculated values:

$$F = 8.15 \bullet 10^{-6} + 2.65 \bullet 10^{-7} + 4.4 \bullet 10^{-6} + 0 \approx 1.3 \bullet 10^{-5}$$

8.7 Coherent and Noncoherent FTA

Maybe this chapter is a good place to explain something else concerning the failure and success while implementing the FTA. The FTA usually works with faults of equipment and failures of operation (activities, processes, etc.), but sometimes a NOT gate might need to be put in the Fault Tree. This can happen if the normal functioning of an equipment or operation needs to be calculated, so the normal functioning is shown by a negation of the fault or failure. Actually, as you can see from my examples, I present the faults or failures with the negations of their normal operations because this is in accordance with the Safety-II approach for incidents and accidents. So, in this book, I am actually working with a noncoherent FTA, even though I am not implementing NOT gates.

The traditional approach states that the NOT gate should be entered into the FTA as a normal operation, and this complicates the situation. One simple example is the air conditioning (A/C) during a fire. In fire situations, the normal functioning of air conditioner will be risk because it pumps oxygen into the room, which supports the fire. There are other types of A/C (let us say when the fan is ON) that can spread the poisonous gasses to other rooms, thus creating hazard for people who are not directly endangered by the fire. In such a situation, the normal functioning of the A/C in the FTA should trigger an alarm to switch the A/C OFF when the fire happens. But, in the kitchen in the MTC campus, where I used to live, the A/C was always ON with the intention to "dissolve" every quantity of gas that may leak due to any problem. So, in this case, the normal functioning of A/C is to set up a barrier to prevent gas ignition in the case of gas leaking.

This type of FTA (with NOT gates) is called noncoherent FTA, compared with the classical one, called coherent.

The noncoherent FTA is a situation when the Failure space and Success space* meet each other. This can create problems with the probability calculations (especially with the software calculations). If this happens, then we are dealing with a poor design; so if you execute the FTA on the designed product, this is a signal that the design should be changed.

In some situations, the XOR gate can also create noncoherent FTA. The XOR gate can be presented as:

$$A \oplus B = A \bullet \overline{B} + \overline{A} \bullet B$$

The XOR gate shows that the result can be a success (1) only if one of the inputs happens. If both are missing or both are present, XOR will result with 0 (event does not happen). So, looking at the formula and diagram (Figure 8.3), I can notice that its use in FTA automatically creates a noncoherent FTA for events A and B.

There are different opinions among the scholars. Some consider that the probabilities of the failures are so small that the negation of the failure is almost 1 and as such, it can be ignored. You can notice from my examples that this is not always the case.

Let us talk about the Piper Alpha accident for example.

It was a disaster that happened in 1988 at the oil platform in the North Sea. One hundred sixty-seven people died and the rig was totally destroyed. The accident was so strong because Piper Alpha was also the transfer point for two additional platforms in the vicinity (Tartan and Claymore) that were

* Paragraph 6.5 (Failure or Success) in this book.

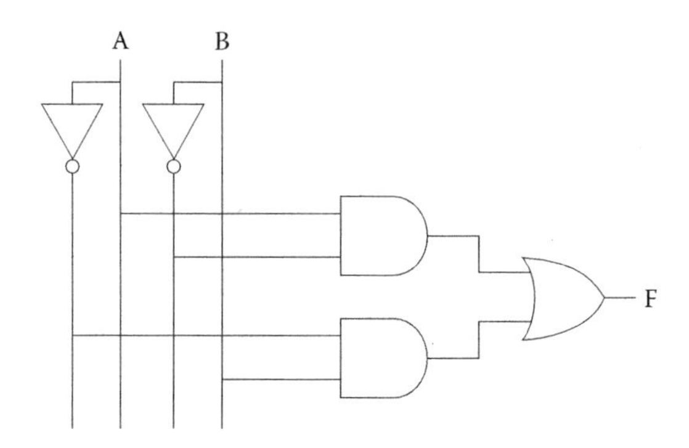

FIGURE 8.3
Diagram of noncoherent FTA for F = A ⊕ B.

supplying Piper Alpha with gas through two pipes of 45 cm diameter each. When the explosion happened, these two rigs did not stop pumping gas, which actually supported the fire on Piper Alpha. The supplying of gas was their normal working operation and if I would like to present it in the FTA, it should be negated and will produce a noncoherent FTA. Anyway, it will help to understand that stopping (negating) the normal operation (supplying gas to Piper Alpha) on Tartan and Claymore will help to control the accident on Piper Alpha. This is good example of the benefit of noncoherent FTA if it was applied in advance.

Another example is my car: Ford Focus. In the car, there is an inertial circuit that will stop the fuel from entering the front side of the chassis* in the case of a crash. This is made to protect the fuel from leaking, which can cause a fire in the case of a crash. So, this circuit should function normally to cause a defect in the normal functioning of the fuel supply system in the case of a crash and should fail to do so if no crash happens. There is no other FTA presentation for this circuit except for the one with the noncoherent FTA.

In the previous cases it does not matter if I am speaking about failure or success: both FTA must be noncoherent!

Finally, noncoherent FTA must be used in aviation, where every piece of ground or aircraft equipment is doubled and associated with a particular monitor. Simply, like in my example for the LLZ, both monitors must register a failure of operating of the first transmitter and must change the operation to the second transmitter to provide a normal operation of the LLZ. The mixture of failures and successes is evident.

Generally, dividing FTA into coherent and noncoherent makes sense for theoretical and scientific purposes. Considering that the scientific context and the industry context differ a lot, not using noncoherent FTA in industry

* The truth is that a lot of manufacturers are using this device in their cars, so it is not only Ford.

and real life limits the possibilities of the overall analysis. In the beginning, I explained that the configurations that I got through the FTA are similar to the applications executed by computers. Such limitations do not exist for computers. In the computer processors, there are thousands of NOT gates. In addition, the FTA by its ontology is a mathematical tool that uses the Boolean algebra, and mathematics has no such limitations too.

Speaking about failures and successes in paragraph 6.5, I stated that they complement each other. Even the probability formula of each event is twofold: it will succeed or it will fail. So, failure is the negation of success and success is the negation of failure. If I use one event (process, equipment, activity, etc.), I can produce two trees: Fault tree and success tree. Even in these, negations of the negations will exist.

I disagree with the scholars who say that you should not use noncoherent FTA in industry, and I do believe that (considering the present situation in industry, where the systems are very complex) using noncoherent FTA is a necessity.

9

FTA for My Home Fire[*]

9.1 Determining the System and Its Boundaries

To give another example of how I conduct the FTA, I use the example of building a Fault Tree for a fire breakout in my home (home fire). Even though I have a good understanding of how my home functions, for the sake of teaching, I will explain everything that matters. I will only deal with fires that can start in my home because this is place where I have control. Fires that start outside my home will not be considered. This does not mean that they will not endanger my home, but it is different type of protection that I need to implement and not part of this chapter. So, the external boundary of my system in the FTA is the interior of my home.

Determining the internal boundaries, I stick to things in my home that can produce fire. This means that I will only consider things that are part of my home and can produce fire due to improper use. For example, using my vacuum cleaner that has a damaged cable can produce a fire if the wires touch each other and produce a short circuit. The sparks from the huge current can touch my carpet and start a fire. So, the resolution of my FTA will be the vacuum cleaner only. I will not consider the parts inside my vacuum cleaner and their contribution (damaged cable!) to the caused fire. It will make the FT simpler and it will fit my purpose of building an FT for my home fire.

In addition, I will not consider the electrical installation in my home as a possible reason for the fire. It is embedded in the walls that are made by bricks and concrete, so even if the fuses fail, only the cables (wires) will be burned and no fire can be produced. Even though a lot of fires start because of improper use of high-power lighting or usage of extension cords,[†] this cannot happen in my house. The reasons are as follows:

[*] Most of the data about fires used in this chapter are reprinted with permission from the NFPA's online report, "Fire Causes by Month," Copyright © 2014, National Fire Protection Association, USA. I am very grateful that they allowed me to use their data for the purpose of this book.

[†] By NFPA, 46% of all fires for every electronic equipment is started from cables and cords (statistics for the period from 2006 to 2010).

The installation of the lights in my apartment is made for light bulbs from 40 to 100 W, but I am using light bulbs that are compact fluorescent (energy saving!). The biggest electrical consumption of this light bulb is 28 W (equivalent to the light produced by 100 W ordinary incandescent light bulbs). So, they cannot overload the electrical installation* in my home.

The extension cords that are used in my home are made by me (I am a Graduated Engineer of Electronics and Telecommunications!) and all of them are produced with higher standards (thicker wires in the cables!) than the electrical installation cables used in walls. If the load of the cords is too large, then the electrical installation will be damaged first (if the fuses do not react!) and the extension cords (made by me) will not be damaged. They cannot generate high heat and therefore they are unable to produce a fire.

9.2 Gathering Knowledge for Fires

To gather knowledge about home fires, I went to the best place in the world: the Internet! I went on Google and experimented with a few titles for home fires: elements of fire, home fires, statistics for fires, and so on. I do believe that almost 6 MB downloaded material from the Internet was enough to start with the building of a general Fault Tree (FT) for home fires. The general FT is presented in Figure 9.1. Note that only primary events are presented. Going into details at this stage will cause additional confusion and I do not like that. Later, when specifically dealing with my home, I will go into details because then that will be necessary to analyze an FT. Anyway, all these primary events could be separately investigated as Main Events for the particular Sub-Fault Trees (SFTs).

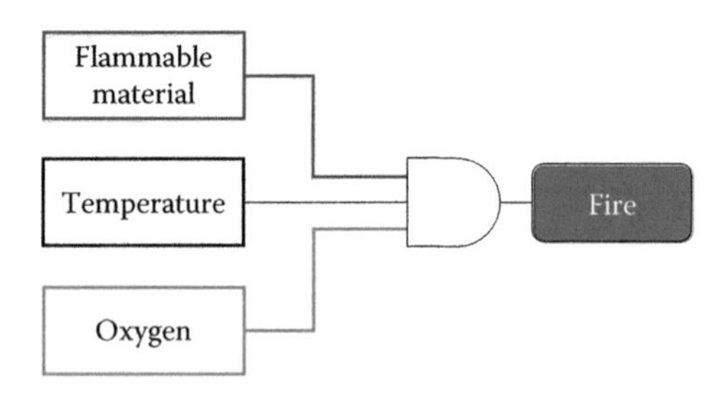

FIGURE 9.1
FT for a fire.

* Overload of electrical installation is caused by improper load, which produces huge current in the wires causing them to heat and even melt. If the wires (cables) are thick enough, the current will be absorbed by the wires and it will not produce heat.

As I have mentioned in paragraph 6.7 (Effects, Modes, and Mechanisms), three elements are needed to start a fire: material that can burn (***matter***), ***temperature*** (ignition), and ***oxygen***. So, the matter is a flammable material, usually wood, paper, textile, plastics, or something else. The temperature provides ignition of the flammable material, but without a high temperature, the flammable material and oxygen can coexist without any danger. So, the general diagram for fire should look as the one in Figure 9.1.

Figure 9.2 shows the general FT for a home fire and you can notice that the primary events are connected with the OR gate. The reason for that is that only one event is enough to start a fire and all these events are independent. These events could produce ignition in the form of heat or sparks that will provide a high enough temperature to start the fire.

But this is not enough…

For my general FT, I need to add the material that can burn (flammable materials as textile, wood, paper, etc.) and oxygen to the diagram. The oxygen is present in any home, so it is added directly to the AND gate. All other flammable materials are added to the AND gate too, but through an OR gate (in accordance with Figure 9.1).

Explaining all primary events in Figure 9.2, I use the statistics from a document* called "Home structure fires" prepared by Marty Ahrens and published in September 2015 by the National Fire Protection Association in USA. There are other documents on the Internet, but this one seemed superior to all others because of its high quality of data.

Let us explain the primary events in Figure 9.2.

Electrical equipment fires are caused by a malfunction of the electrical equipment in the house. I can divide these equipment into two categories.

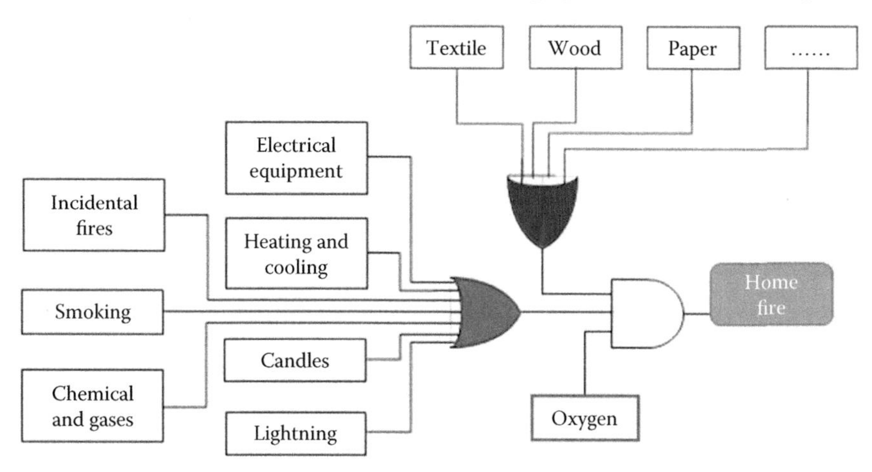

FIGURE 9.2
FT for a home fire.

* Downloaded from www.nfpa.org.

The first category includes home appliances consisting of cooking appliances, fridges, food processors, washing machines and dryers, air conditioners, fans, and more. The second category includes electronic equipment consisting of TV sets, radios, music boxes, recorders, cameras, computers, electronic clocks, weather stations, and so on.

Heating and cooling are air-conditioning equipment modes necessary to keep homes at a particular temperature (provide heating during winter and cooling during summer). Such pieces of equipment are heaters, ovens, fireplaces, air conditioners, fans, and so on.

Candles are often used in homes during particular events (celebrations, birthdays, romance, etc.). It is interesting that for the period from 2009 to 2013,* 3% of all fires in the USA were caused by candles and 3% of all the human casualties in the fires were from fires started by candles.

Surprisingly, lightning is also a contributor for home fires! Statistics of NFPA showed that from 2007 to 2011 in USA, 19% of all fires caused by lightning during storms happened to houses or human building structures.

Incidental fires are unintentional fires that happen during our regular home activities. This can happen when cooking, cleaning, repairing, heating, partying, and so on.

Smoking is also a type of unintentional fire, but due to its high contribution to home fires, it is put in its own individual category. This happens due to improper use (misuse) of cigarettes or due to improper extinguishing of finished cigarettes.

Chemicals and gases also contribute to home fires mainly because they are not kept in the right places and a fault in electrical installation in combination with them is very dangerous. Sometimes, it incidentally happens that different chemicals mix with each other, starting a chain reaction that will result in a high temperature or even fire. Gases contribute to fires due to their use in cooking and heating activities in homes.

As we can notice, there are situations that can be categorized in two or more of these categories, but this is not important for our purpose.

9.3 Fault Tree for My Home Fire

To build an FT for my home, I use the diagram in Figure 9.2, but I adapt it to the specifics of my home. The diagram of the FT for My Home fire is presented in Figure 9.3. Let us explain.

Smoking and lightning are nonexistent in My Home fire diagram.

* *Source*: "U.S. Home structure fires fact sheet," published by National Fire Protection Association (www.nfpa.org).

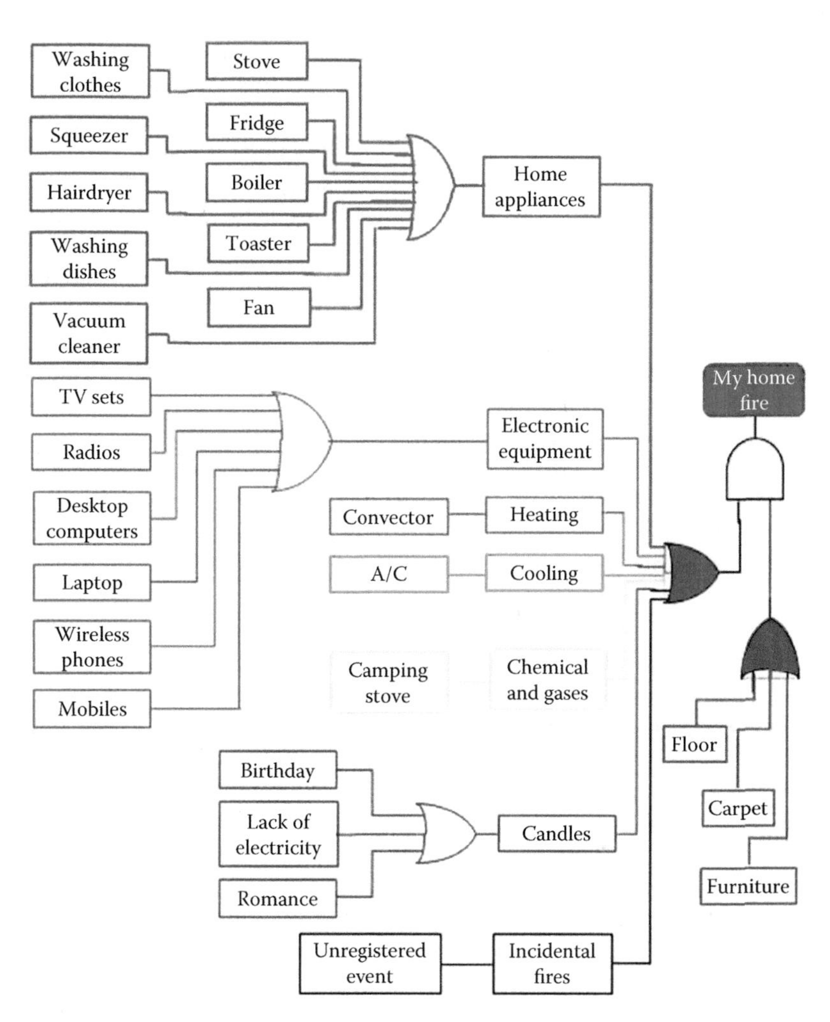

FIGURE 9.3
FT for My Home fire.

Smoking as a cause of fire is removed totally because no one in my family (me, my wife, and my children) smokes. Even the guests respect our *non-smoking environment* and do not smoke inside. Even though we tell our friends that it is okay for them to smoke inside, they still go to the balcony. When me and my wife are not at home, my children's friends are often at our place and they also smoke in the balcony. And there is nothing in the balcony that can be burned: it is made by concrete, has a metal fence, and the door and windows are made by PVC (polyvinyl chloride), which is fire retardant. So, in my home there is no possibility that a fire can outburst due to smoking.

Lightning is not on the list because my home is in the city, at the second floor of a three-floor building with many buildings around. Scientifically, this is not a place where lightning can strike! The danger of lightning falls on lonely houses, outside cities, in the plains, and usually with no trees around. In addition, there are higher buildings in my neighborhood and the thunder protection is good for all of them. Also, lightning is factor that is outside My Home boundaries.

As you can see, I removed the oxygen from the FT. There is no need to have it in mind knowing it is always present, which means I cannot deal with it. If there was no oxygen in my home—there would be no life at all!

Another thing that can be noticed in the diagram in Figure 9.3 is that I have divided the Electrical Equipment from Figure 9.2 into two parts: Home Appliances and Electronic Equipment.

Home Appliances are things in my home that support our essential living needs (lights, food and drink storage, preparation, consumption, etc.). Such things in my home are: one electrical stove, one vacuum cleaner, one fridge (freezing and normal cooling), two water boilers (60 and 10 L), one washing machine for clothes and one washing machine for dishes, two electrical squeezers (for preparing fruit and vegetable juices), one toaster, one hairdryer, and two ventilation fans (in the bathroom and toilet). As you can notice, all of them are connected with red lines and OR gates (common gate for Home Appliances fires), because if only one of them malfunctions, it is enough to start a fire. Of course, not all type of malfunction can cause a fire!

Electronic equipment is equipment that is used for other purposes to make my life easier (and more fun!). These are: four TV sets, two radios, two desktop computers, four laptops, three wireless phones, and four mobile phones. Last three (laptops, wireless, and mobile phones) are fire hazards due to the charging devices used for them. Also, anyone (and only one!) of them is enough to start a fire, so all of them are inputs into the blue line OR gate (common gate for Electronic Equipment fires).

The last thing that can be noticed in the diagram in Figure 9.3 is that I have divided the Heating and Cooling from Figure 9.2 into two parts: Heating and Cooling. The reason for that is that I have seven electrical heaters and four air conditioners for cooling. They are independent, so I need to consider them separately. Heating and cooling do not need any logic gates, meaning that I can connect them directly to the main (filled with blue color) OR gate.

The same situation is with chemicals and gasses. I do not have and do not keep any flammable chemicals, but I have a camping gas stove, which I rarely use to prepare coffee or tea. So, it can be directly connected to the main blue filled OR gate.

Candles are only used on three occasions, but not very often. First one is during birthdays, but now that my children are all grown up we do not prepare birthday cakes with candles. Lack of electricity does not happen very often, but a traditional Macedonian habit is that every house needs at least two candles in the case of emergency situations. The third one is for romantic purposes, and I think there is no need to explain that.

Incidental fires are fires that can happen in strange ways that were not registered until now. Fire scenarios can be extremely variable and (as an old friend of mine used to say) the Murphy's Law is highly applicable for such situations. In paragraph 2.2 ("Black Swan" events) and 2.5 (Basics of probability) I already mentioned cases which have not been registered, yet they still happened (9/11 and Fukushima).

So, these are the main causes of fire in my apartment, but this does not mean that if one of them happens a fire would outburst. Any one of these causes (by itself) is not enough. All of them can produce heat or sparks (temperature!), but a fire can only start if the sparks (or heat) touch a flammable material. Such flammable materials in my home are carpets, floors (wooden parquet), and furniture (textile and wood).

That is the reason why any of these materials are connected as inputs with a violet OR gate, which (together with the blue OR gate) is connected to an AND gate. Both of the outputs of the blue (heat or sparks) and violet (flammable materials) OR gates must happen at the same time (AND gate), so they can, together with the ever-present oxygen, trigger a fire.

9.4 Qualitative Analysis for My FT

The next step will be to check my FT again and give it to my wife and children for comments. They are living in same apartment with me, so although I have the best intentions, maybe I have forgotten or missed something. When I showed them the diagram in Figure 9.3, they made fun of me! They were looking at me with a strange face, but when I explained them the purpose (writing a book!), they became more serious. Even though they put some efforts, I was praised because I took almost everything into consideration. So, no comments or advices from their side!

The next step is to produce an empirical Boolean function for this FT, but I will not do it simply because there is no reason for it due to the nature of the example.

First, all of the inputs in every OR gate are independent events that are inputted in only one OR gate. It means that there are no possible combinations that can arise from these inputs, so they will appear in the empirical Boolean formula only once. Having each of them only once, I will not be able to do any kind of simplification.

Second, the modularization is already done, but I will not present the SFTs for each one of them because there is no reason for that. The purpose is to explain how to build the FT and this is enough to understand the process.

Building all other SFTs will only complicate the explanation and will not bring us any benefit.

So, the qualitative analysis for the FT of My Home fire is finished.

9.5 Quantitative Analysis for My FT

Quantitative analysis means to dedicate particular frequencies or probabilities to the events in my FT and use them to calculate the overall probability (or frequency) of a fire happening in my home. The formulas* that arose from my home fire FT are:

For home appliances:

$$P(HA) = P(ST) + P(F) + P(B) + P(T) + P(F) + P(WC) + P(WD) + P(SQ) + P(VC) + P(HD)$$

For electronic equipment:

$$P(EE) = P(TV) + P(R) + P(DT) + P(LT) + P(WP) + P(MP)$$

For heating:

$$P(HE) = P(C)$$

For cooling:

$$P(CO) = P(AC)$$

For chemicals and gases:

$$P(CG) = P(CS)$$

For candles:

$$P(CA) = P(BD) + P(LE) + P(R)$$

For incidental events:

$$P(IE) = P(UE)$$

For flammable materials:

$$P(FM) = P(CA) + P(FL) + P(FU)$$

* In every formula, P stands for probability and the capital letters inside the brackets are first letters of the name of the equipment.

Connecting them into a total formula we get:

$$P(MHF) = \left[P(HA) + P(EE) + P(HE) + P(CO) + P(CG) + PCA)\right] \bullet \left[P(CA) + P(FL) + P(FU)\right]$$

But calculating this is a problem, as you can notice I cannot simplify the formulas. In addition, the biggest problem here is to find the data for all these probabilities mentioned in the formulas.

I tried to have a scientific approach to this part and contacted companies that have produced the things in my home (the primary events). I got the needed information by contacting them directly (through e-mail) and by downloading different data sheets from their websites. Mechanical things cannot generate heat if there is no extensive friction between moving parts. That is the reason why I was only searching for the reliability of the electrical and electronic things which I am using in my home. Electrical and electronic things use voltage, which (if it is shortcut) can produce sparks.* Normal voltage in the appliances produces current inside and this current depends on the internal resistance. If the wires and internal resistances are not adjusted to normal voltage, then it can produce a large current that will generate a high temperature (heat) inside. But, these things are already known for a long time and today, the protection (for cases like this one) is extremely good. Therefore, the chances of having a fire produced by electrical and electronic equipment (if used properly!) are equal to 0.

But looking for data regarding reliability, the result was that all of them have very high reliabilities regarding faults if they are used in accordance with the product specifications. Even though reliability is about the faults of the equipment, it does not mean that every fault in electrical or electronic equipment will result in extensive heat or sparks that can produce a fire. Or, as a guy from AEG wrote in his e-mail to me regarding the heating convectors from his company: If they are not covered by something (which is strongly forbidden!), then it is impossible for them to produce a fire! Covering the heating convectors with something is extremely stupid, so if I (or someone from my family) do not do something stupid, the cause of the fire in my home cannot be from any of these electrical and electronic things, but from a stupid behavior of a member of my family.

So, the overall probability of having fire in my home regarding the Home Appliances, Electronic Equipment, Heating and Cooling depends only on their use (misuse) (human error!). So, let us see how the situation is.

I have never experienced a fire in my home! But let us go part by part. Let us speak first about the elements of Home Appliances. For all of the 10 elements there, I have not experienced any problems with 9 of them. Once, I put cup of milk on the stove (no microwaves in that time!) and I forgot it.

---

* Sparks (in this case) are huge amounts of current released in a very short period of time!

After 1 h, it resulted in an extreme odor, which spread to the whole building. Similar thing happened to me when I was teenager, which means that the probability to make such a mistake for myself is once in 26 years! But this is only for me. It is not I, but my wife that is mostly using the Home Appliances and she does not remember that she was even close to cause a fire, ever! My children are not using them very often (except for the microwave oven) and they also do not remember that something that could lead to a fire happened.

Speaking about the Electronic Equipment, it is used more than the Home Appliances and the only thing that makes me worry is that sometimes my older son is charging his mobile phone in his bed while sleeping. This is not good because mobile phones are designed to produce a particular amount of heat and if they are not charged on a place where natural cooling is provided, the chances of fire are evident. Putting the mobile phone in the bed during charging, he can incidentally cover it with blanket or something else and the ventilation will be lost. Actually, this hazard happens very often in my house. Knowing this I usually check the phone of my older son when he is sleeping and if it is in his bed I move it to a place where it can be cooled naturally. All others, they put their phones on places where the cooling is OK.

The Camping Stove (Chemicals and Gases) is used only by me and I am pretty aware of the problems that the gas can produce! I am camping for almost 40 years and as far as I know, I have not made any mistake. I was always pretty cautious dealing with it! So, I find this probability to be equal to 0.

Candles are used extremely rarely in my home, but the situations when they are used are characterized by a low awareness of the bad things that can happen. Because of this, I always try to warn my children about these hazards and I can only hope that they will remember my words.

Incidental Fires are Unregistered Events for which I cannot recognize the mechanism of how they happen. Maybe some of them never have happened. So, for such events (Black Swan!), I cannot dedicate any probability because there is no appropriate data for them and they happen extremely rarely.

Unfortunately, I cannot associate any probability to any part of my FT. The only probability that I can associate could be the one with the cooking in 26 years (the cases when I forgot the milk on the stove). But, this is not reliable data. I cannot find such data because, in the past, fire protection agencies did not calculate the probabilities for fire by using a particular piece of equipment at homes. The statistics was dedicated only to collecting data for the causes, sorting them by the type of the equipment. The reason for that was that the system that analyzed the data about the probabilities of fires was very complex because it included the human behavior. This makes the companies who are producing the electric and electronic equipment focus on these particular cases of causing a fire and they strive to improve their equipment, which will not allow these cases to be repeated. And honestly speaking, they were extremely successful in this: Today, the proper use of every kind of electric or electronic equipment cannot cause harm. Actually, the humans are those who endanger their homes by misusing the equipment. Reading the Operating Manual carefully and following the instructions inside are of utmost importance for achieving home safety.

9.6 "Epitaph" for My Home Fire FT

The example I use here might look strange.

I wrote plenty of pages on how to build and analyze the FT for a fire in my home, but I said nothing. But this was all done on purpose! The main thing is not only to build an FT, but to have enough numbers (data) to put inside the empirical formulas for that same FT. A good FT without good data is useless. And, in my humble opinion, that is the biggest problem in the usage of FTA. I presented an example that is common to everyone, but I cannot produce any outcome of all of these FTs and formulas due to the lack of proper information.

But, in most of the risky industries, the situation is not so bad. There are a lot of statistical analyses done by companies or regulatory bodies and some of them have already produced general probabilities that are applicable for the practical use of the FTA. Anyway, the main point is to make the Safety Managers produce their own databases for the events in their companies. These data combined with the general data from other organizations or bodies can provide a great basis for good calculations. But please understand: The BM comprised by FTA and ETA is just a methodology (tool) for getting a better picture of what is going on and the safety is much more than just numbers.

Anyway, the reason I cannot do any calculation for my home fire is that companies that are dealing with home products (whatever it is!) are all well aware of the capabilities of humans to be unserious during the usage of their products. So, a long time ago, they undertook certain steps to produce good Error Proofing equipment that minimized the possibilities of a fire.

But home fires happen every day, so all of us need to live with this risk. What could the solution be?

You must be serious and aware that the misuse of electrical and electronic equipment and products can cause a fire. So, we need to monitor them all the time. Whenever the risk cannot be decreased to an acceptable level, we must monitor the situation. Of course, monitoring is not enough if we do not have preventive and/or corrective measures, but I will speak about this later.

10

Event Tree Analysis

10.1 Introduction

Event Tree Analysis* (ETA) is a method to analyze the effects (consequences) of the Main Event and to find the right action on how to mitigate it. An inductive method (bottom-up), in BM, the ETA starts when the FTA finishes.

It was developed in the first half of the 1970s by a group of engineers who were doing the risk assessment of the WASH-1400 nuclear reactor. It was used together with the FTA. The FTA helped them find the failures and faults and the ETA helped them consider the further scenarios for the development of the Main Event.

The ETA utilizes a tree structure known as the Event Tree (ET), but it is not in the same form as the FTA. For every consequence, I gradually produce particular mitigations where only two logical states (0 and 1 or Failure and Success) are presented for each mitigation. The ETA is applicable to all consequences of the particular Main Event. Of course, it is best if it is conducted during the designing phase of the system. Similarly to the FTA, it is very important to have excellent knowledge about the functioning of the system. This should not be an issue because the ETA is usually conducted by the same person who is dealing with the FTA, so he or she is already familiar with the system. Knowing the mechanisms, modes, and effects of the FTA, it is extremely helpful to use the ETA to produce scenarios of how the consequences of Main Event will develop later and to find particular solutions for elimination or mitigation.

Let us discuss a little bit about the measures that should be undertaken after the Main Event happens.

I said that a Main Event produces effects. I also said that I can call these effects consequences because they produce unwanted situations for us. I can deal with these consequences in two ways: eliminate them or mitigate them. I can also ignore them, but can I really afford that? The answer is yes, if they do not cause any damage to human and assets, I can afford this! But is it

* Even though I use the ETA as a Post-Event analysis in the BM, it is a method that can be also used as a Pre-Event analysis.

worth thinking about? If I can ignore the consequences, why do I really consider the Main Event? Ignoring the consequences of the Main Event tells me that something is elementarily wrong with my system!

If you look for the word "elimination" in dictionaries, you will see that it is defined as to reduce or decrease something to the level of disappearing. So, to eliminate something means to make it non-existent. This is the best solution that I can apply to unwanted events or consequences. The problem is that it is not always possible to do it immediately. I can eliminate the consequences, but this needs time and money. It is easy to find money, but time is a problem! A building destroyed in an earthquake can be built again, but not immediately and for a bigger price than before. To rebuild a seven-storey building, you need approximately 14–18 months and a bigger price than before because you would need money to clean the leftovers of the old building too.

Considering that it is not always possible to eliminate the consequences, in a case like that, effort must be made to mitigate them. Destroyed buildings due to earthquakes equals to no homes for the people, so we must provide to them temporary homes (usually tents or public sport halls), which is one of the first mitigations.

Mitigation is defined in dictionaries as an action to reduce or decrease something to an acceptable level of coexistence. It is not perfect, but I can live with that. The nature of mitigation can be different. It can be some kind of a barrier that will stop and/or isolate the scenario of development of the consequences, or it can be an activity that will recover the normal situations. Sometimes, mitigation just buys us time to deal with the consequences. I should think about it as a system of contingency plans for emergency situations that should be produced in the ETA in advance. This means that I should know how to apply the mitigation before the Main Events happens. This is the reason why the ETA should be applied in advance! In such cases, the ETA is very helpful in preparing us for the worst!

Statistics show that after a plane crash, there are many casualties, although sometimes, some of the passengers survive the crash. Anyway, due to the late arrival of the SAR (Search and Rescue) Teams on the crash site, medical help was not offered immediately and they died too. So, having trained SAR Teams ready can make a difference between life and death in the case of an aircraft accident.

10.2 Main Event and Consequences

After the Main Event happens (which means: I cannot eliminate it anymore), it produces different consequences that I need to eliminate or minimize. I said consequences (plural) because there is never only one consequence. If I have a car crash, the consequences can be different: medical (level of injury, disability, etc.), economical (car repair, damage on assets, less money due to sick leave, etc.), legal (prosecution if guilty, damage liability, etc.), and so on. Whatever the consequences, everybody would like to mitigate them. And ETA is very helpful in doing this!

Today, I can use the ETA for two purposes. The first one is to find how the Main Event will develop (or to find possible scenarios) if it is not stopped immediately. The development of these scenarios (how the consequences of Main Event progress) is strongly dependent on the data available about the Main Event and the environment around it. If I am speaking about a Main Event which has already happened in the past, then there are already some data about its development. But if the Main Event has not happened in the past, then I should assume its development or I should do a modeling simulation. There is no qualitative difference between making an assumption and modeling the scenarios for the development of the consequences. Both are used due to lack of data and present our understanding (perception of the Main Event), which means that both are highly subjective. Assumptions are cheaper, but modeling simulations are more popular. I would honestly stick to assumptions because modeling human behavior in incidents or accidents is still far away from reliable.

Knowledge about the scenarios (how the consequences progress) brings us to the second purpose of the ETA, which is to use this knowledge to implement mitigation measures to deal with consequences. This purpose prevails when the ETA is used in the BM. Actually, the ETA is about building a mitigation plan based on the scenarios determined by the knowledge or assumptions about the Main Event.

There is one more thing that may, and should, be used in ETA: including the measures to provide a fast recovery of the operations. The ETA becomes more complicated with those recovery measures, but sometimes those measures can be integrated into the mitigation measures or even in the system before the FTA is done. This integration highly improves the effectiveness and efficiency of the ETA and brings the company additional economic relieves. I recommend this as a good thing to do, but in the end, it is your choice whether you will do it.

The ETA is conducted with the intention to mitigate and control the consequences, stopping them from producing long-lasting damages. For example, if there is fire in an apartment, the first thing that firefighters do is to put the fire in control (by stopping the fire from spreading to other apartments in the building) and then proceed with stopping the fire (mitigate the damage).

10.3 "Black Swan" Events and ETA

The Black Swan events are mentioned in Chapter 2, paragraph 2.2. These are events that happen as a total surprise where no one could have predicted such a situation. There are no data for them and they seem almost impossible to happen or to be predicted. These are events so strong that the consequences are terrible. This can affect our ETA.

Maybe it sounds strange, but Black Swan events are easier to be handled in not-so-safety organized companies than in companies with a good safety culture and training. Reason for this is that in companies with good safety organization, such events will produce a complete blackout between the employees. They are trained on safety procedures and mitigation measures for almost every hazard and having in mind that the Black Swan events are not anticipated, there is no procedure on how to fight it. So, the happening of a Black Swan in such companies will leave the employees without a timely response.

On the other hand, the not-so-safety organized companies handle safety by improvising. So, when something bad happens, the employees know that there is no procedure anyway. Having in mind that they are already good in improvising, it is easy for them to try and improvise something to handle the consequences. For them, it will be like the Main Event was not a Black Swan Event, but a normal one.

Anyway, there is no remedy for stopping Black Swan events. During the FTA, you may be more innovative to anticipate such an event and make your system resilient to it, but this is highly improbable.

There is no remedy for stopping them, but dealing with the consequences with ETA should be easy. That what is not predictable is the failure modes of how the Black Swan events may happen, but, the consequences should not be very different from those which are already known. The history of such events tells us that they can differ by volume (they are usually very huge events!), but not by nature. So, if employees do not spend much time on how the event happened and focus only on the consequences, they may be able to fight Black Swan events. Simply, they will use the already established elimination, mitigation, or recovery measures for the same consequences of different events. So, the ETA must be more open and free minded for execution.

10.4 Executing ETA

There are a few steps in conducting the ETA. Figure 10.1 shows a general flow chart of these steps.

For each Main Event, there are plenty of consequences.* A list with all of them should be produced and an Event Tree (ET) needs to be produced for each of them. To be good at producing the ET, I need to have a good knowledge of all consequences that will help me predict all the possible scenarios

* Some of the authors in this area called the list of consequences a ***Consequence Spectrum***, and I do not have any objections about that!

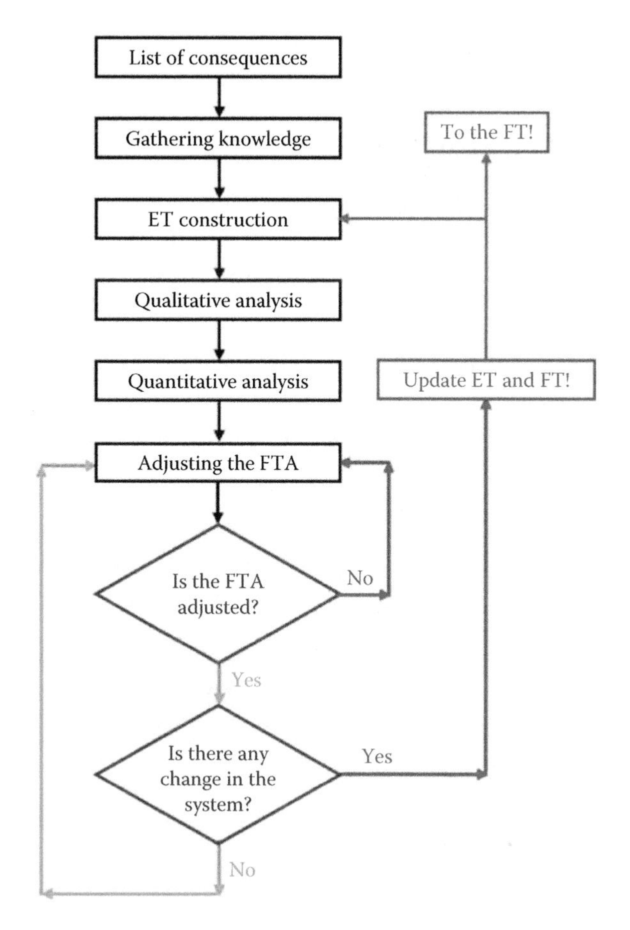

FIGURE 10.1
Flow chart for executing ETA.

of how each consequence may develop. Actually, mitigation measures are responses of different scenarios for each consequence. Sometimes, same consequences can develop different scenarios. You should also take this into consideration and produce an ETA for each scenario! Without these things, a reliable ET construction is not possible.

I will construct the ET for each consequence and execute a full ETA on each ET.

The execution is made by performing a qualitative and quantitative analysis of the ET. The analysis will show me if there is any need to change something in the FTA because some of the scenarios can give me hints on how to make the mitigation easier by changing something in the system before the Main Event happens. Do not forget, FTA deals with the system that produces the Main Event, whereas ETA deals with the consequences caused by this Main Event. This interaction between the FTA and ETA is actually an adjustment of the normal functioning of my system, which could later help me mitigate the possible consequences if a Main Event happens.

This is one of the most important benefits that characterize the BM. Considering that the BM is holistic, it deals with the overall situation regarding the Main Event (Pre-Event analysis (FTA) and Post-Event analysis (ETA)). This means that BM is giving me an opportunity to improve things by intervening where and when I see fit: before the Main Event (in FTA) or after the Main Event (in ETA). And, of course, if I treat the Main Event as a black box with input (FTA) and output (ETA), an old engineering saying* fits: You need less efforts and resources to achieve something during the output (ETA) if you intervene during the input (FTA).†

This means that making the right change in the FTA will make it easier to improve things and mitigate the consequences in the ETA. But this triggers another activity: If the FTA is changed in this step, I need to check the whole FTA again (with the change implemented!).

I need to analyze how these changes in the system affect our FTA and ETA. So, I will check again if there is any change in the system that is not imported in my FTA and ETA. If there is, I will repeat the analysis, and if there are not, my analysis is finished.

As you can notice, executing the ETA is an iterative process. Any change due to mitigation shall be considered as a change in the system that affects not only the FTA but also the ETA. Of course, it depends on the change.

10.5 Event Tree (ET) Constructions

When the ETA is used to eliminate or mitigate consequences of Main Events that are not analyzed in advance, then you need to gather knowledge about the systems where the Main Event is produced. For the ETA as part of the BM, I already know the system under analysis (FTA is finished!), so this situation is clear. For such cases, the procedure of constructing the ET is simple:

1. Identify all of the consequences (scenarios or sequences of events for the development).
2. Determine the mitigations for every consequence.
3. Find the probabilities of Success (S) and Failure (F) for every mitigation.
4. Calculate the probabilities of success or failure for each scenario for the mitigation of each consequence.

* See paragraph 12.5 (Control by Feedback).

† More about this in paragraph 12.2.

The ETA requires the same knowledge, experience, and skills as the FTA. Similar to the FTA, after the ET is finished, you need to share it with other employees and ask for their comment, advice, or suggestion. It can significantly improve the analysis and can produce some other ideas for mitigation.

10.6 How the "Ideal" ETA Works

The ETA is a binominal method and every mitigation step produces only two solutions*: Success (S) or Failure (F). A diagram that explains the working of an "ideal" ETA is presented in Figure 10.2. I call this ETA "ideal," as I can mitigate the consequence with every single attempt, which does not always happen in reality.

As you can notice, the Main Event has happened and there are a few consequences marked with C1, C2, C3, C4, …, Cn. These consequences are actually different scenarios of how the event will (can) develop after it has happened. I need to try to eliminate every one of these scenarios, and I need to mitigate them through particular actions if I am not able to do so. Due to its simplicity, the diagram in Figure 10.2 only shows the ETA for C1. As you can notice, the first thing I do is M1 (Mitigation 1). When applied, it can have only two

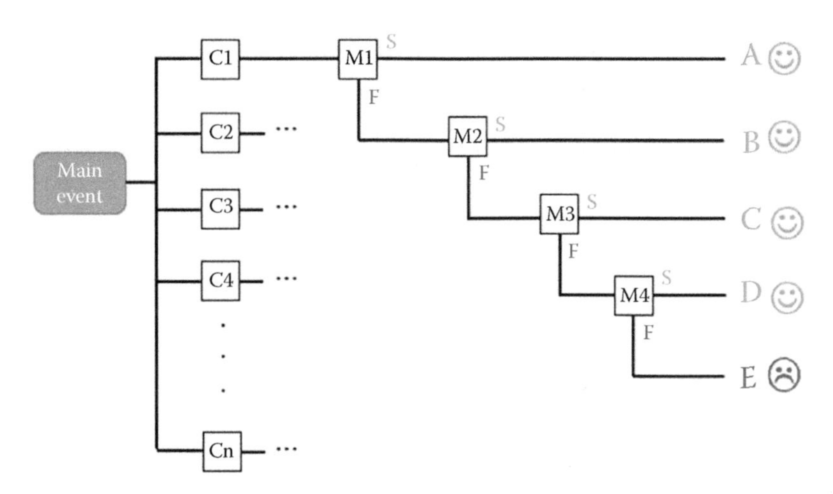

FIGURE 10.2
"Ideal" Event Tree Analysis.

* Some authors use more than two solutions (Partial Failure or Partial Success) and consider that the Partial Failure and Partial Success are too subjective for an accurate differentiation from the Failure and Success, I found it more appropriate to deal with Failure and Success only!

outcomes: Success (S) or Failure (F). If success is achieved, then everything is OK (I can relax!). But, if it fails, I need to find another mitigation procedure (M2). Then, the same applies for M2: There can only be a Success or a Failure. Going further, the same thing will happen for both M3 and M4. I can notice that it is an iterative process until I spend all of my mitigation procedures. So, all of the mitigation procedures produce different outcomes and I like all of them except the last one. In Figure 10.2, all outcomes (A, B, C, and D) are good (Success), except the last outcome E: It is a failure so I would not like it if it happened!

The main point here is that not all A, B, C, and D will be total successes. Some damage of the equipment and/or some injuries will still exist as a result of the Main Event, but all these mitigations (M1, M2, M3, and M4) directing to A, B, C, and D will produce some containment, mitigation, or improvement of consequence outcomes. Outcome E, on the other hand, means that the catastrophic outcome is not prevented, contained, or controlled.

Let us give an example:

If I am ill I will try to "fix" the illness by myself by wearing warm clothes, drinking hot teas, and taking a rest (Mitigation 1). If this does not work, I will go to my general practitioner (Mitigation 2). If there are no results from his therapy, he will probably send me to a specialist (Mitigation 3). If he is not able to solve the problem either, it is highly probable that I will need surgery, so he will send me to a surgeon (Mitigation 4). If this is not working then...

Let us go back to other consequences. Not all of them can be mitigated, but even if they can, the same mitigation would not always work for all of them. Consequently, not all of them will have the same Event Tree (ET). Generally, a bigger ET means that more mitigation procedures are implemented. But even a big ET will not be able to always mitigate all consequences. In addition, not every normal operation (before a Main Event happens) can be recovered without extensive efforts and costs.

10.7 Probabilities with "Ideal" ETA

Whatever the mitigation implemented, there is a particular probability that it will work (produce success). It is important to mention here that the sum of the probability of Success and the probability of Failure is equal to 1, relatively for every mitigation. Of course, this is correct only if the events are not dependent; however, considering that I said that I will assume that,* I will have:

* Look at the end of paragraph 4.2 (Boolean symbols for engineering) in this book!

$$P_n(S) + P_n(F) = 1$$

where:
n is number of mitigation
S is Success
F is Failure

Let us be a bit more specific about this. For M1, sum of the probability of Success of mitigation M1 and the probability of Failure of mitigation M1 are equal to 1.* But, if the M1 fails, I need to implement the M2 measure, which can have two outcomes: Success and Failure. The adding of $P_2(S)$ and $P_2(F)$ should also be equal to 1, but only relatively. In absolute values, I have:

$$P_2(S) + P_2(F) = P_1(F)$$

The same thing will happen with M3, so relatively, if we add $P_3(S)$ and $P_3(F)$, the result will be equal to 1, but absolutely:

$$P_3(S) + P_3(F) = P_2(F)$$

So, generally, I can write that for my case from Figure 10.2:

$$\boxed{P_1(F)} = P_2(S) + P_3(S) + P_4(S) + P_4(F)$$

which brings me to:

$$P_1(S) + \boxed{P_2(S) + P_3(S) + P_4(S) + P_4(F)} = 1$$

But this is not important for me! I would like to calculate the probability of E to happen because it is the event that I do not want to happen (worst-case scenario!). And, following the diagram in Figure 9.1, I can write:

$$P(E) = P_1(F) \bullet P_2(F) \bullet P_3(F) \bullet P_4(F)$$

Looking at the formula for P(E), I can see that all four mitigations (M1–M4) need to fail† for the event E to happen. But, to find the probability of E not happening, I can write:

$$P(\bar{E}) = P_1(F) \bullet P_2(F) \bullet P_3(F) \bullet P_4(S)$$

* If I am working with a "coherent" ETA, then I can always write $P(F) = 1 - P(S)$.
† This is an AND Boolean function (all of them need to happen at the same or one by one or following the order).

This formula presents the probability of E not happening only if M1, M2, and M3 mitigations have already failed. If any one of them succeeds, the probability of E does not matter.

If I want to be scientifically correct, then I should consider that $P_2(S)$ and $P_2(F)$ are conditional probabilities of $P_1(F)$. The reason why I am doing this is because M1 must fail for me to have a chance to implement M2. But, in reality, this does not happen very often. With the "ideal" ETA I am usually choosing the mitigations unconditionally, expecting that under those circumstances this mitigation is the best one. Choosing this mitigation, I am investigating why it could not work and for such a case I am choosing the next mitigation which is obviously quite different than previous one. So usually, there is no correlation between any two mitigations.

Another point is that for the case from Figure 10.2, the following formula will always be valid:

$$P(A)+P(B)+P(C)+P(D)+P(E)=1$$

10.8 How the "Real" ETA Works

In the previous paragraph, I explained how the "ideal" ETA works. I say "ideal" because this is the case where if mitigation M1 is successful then there is no need for additional mitigations. And this is only possible in a "perfect world": One measure and problem solved!

But, in reality, you cannot really be that successful. Usually, you have to work with a series of mitigations (applied step by step, each in its own time) that can eventually provide success. The first mitigation for an aircraft crash is to apply the SAR operation. The SAR Team will provide immediate help to the injured and get a clear picture of the situation (second mitigation). After that the SAR Team must organize transport for each of the survivors to hospitals (third mitigation). When they arrive at the hospitals, an additional (more thorough) medical investigation must be done (fourth mitigation) and the emergency cases will be treated immediately (fifth mitigation). After that, a particular therapy and control should be applied (sixth mitigation). So, a total recovery is possible (not always), but after a particular amount of time, which is mandatory for all mitigation measures to be applied. Anyway, all of these six mitigations have two outcomes: Success (S) and Failure (F). So, considering this, the "real" ETA will look as shown in Figure 10.3.

Using the information from Figure 10.3, I can calculate the probability of all outcomes (A to H):

$$P(A)=P_1(S)\bullet P_2(S)\bullet P_4(S)$$

$$P(B)=P_1(S)\bullet P_2(S)\bullet P_4(F)$$

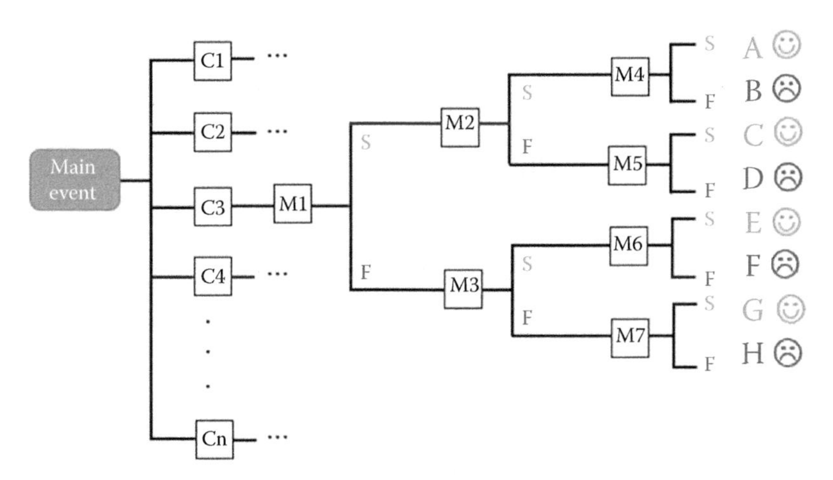

FIGURE 10.3
Real Event Tree Analysis.

$$P(C) = P_1(S) \bullet P_2(F) \bullet P_5(S)$$

$$P(D) = P_1(S) \bullet P_2(F) \bullet P_5(F)$$

$$P(E) = P_1(F) \bullet P_3(S) \bullet P_6(S)$$

$$P(F) = P_1(F) \bullet P_3(S) \bullet P_6(F)$$

$$P(G) = P_1(F) \bullet P_3(F) \bullet P_7(S)$$

$$P(H) = P_1(F) \bullet P_3(F) \bullet P_7(F)$$

Of course, looking for success only, A, C, E, and G are successes. It does not matter that A is a total success and C, E, and G have one failure each. But what is important is the final result because it is realistic to assume that not all mitigation will finish with a success. On the other hand, B, D, F, and H are failures even though they may have experienced some success: The final result is failure. The main thing is to continue to look for other solutions and never give up. Of course, the ALARP principle is also applicable in this area.

Another important point that deserves to be mentioned here again is that always:

$$P(A) + P(B) + P(C) + P(D) + P(E) + P(F) + P(F) + P(G) + P(H) = 1$$

Again, I should mention that these are all conditional probabilities. I already spoke about that in paragraph 10.4 (how the "ideal" ETA works), mentioning that in reality with "ideal" ETA, the probabilities do not dependent on each other. With "real life" ETA, if I deal with a success, then the next mitigation will be dependent on the previous one. This happens because all of the mitigations that are in a row are actually steps of a same chain of mitigations. Also, if M1 fails, I will investigate why it failed using the "real life" ETA. This way I will collect more knowledge about the development of the consequence. This will help me choose an appropriate next mitigation (M3) that will actually be a result of my investigation. So, I can say that there is some type of dependence here. So, probability formula for E should be:

$$P(E) = P_1(F) \bullet P_{3|1F}(S) \bullet P_{6|3S}(S)$$

where:

$P(E)$ is the probability of E happening
$P_1(F)$ is the probability of M1 failing
$P_{3|1F}(S)$ is the probability of M3 succeeding if M1 failed
$P_{6|3S}(S)$ is the probability of M6 succeeding if M3 already succeeded

The other formulas for the outcomes should be similarly arranged.

The dependent events could be a problem here. Sometimes (if there is a need to be precise!), I will treat the mitigations M3 and M6 as Main Events and I will conduct an FTA on them to see how they depend on the failure of M1 and M3. This complicates the situation because now the ETA depends on the FTA for these dependent events. Anyway, as I already explained at the end of paragraph 4.2 (Boolean symbols for engineering, speaking about the OR gate), working with independent events gives you more conservative results and brings you to the safe side. If you can achieve success using independent events in the analysis, then by using dependent events, you will satisfy the safety requirements better.

10.9 Qualitative and Quantitative Analysis

When the Event Tree is fully constructed, I need to check it again if it is correct. This is the qualitative analysis where all mitigations and connections are checked again.

When the qualitative analysis shows that the Event Tree is a realistic model of the consequences of the Main Event, then I may associate the particular probabilities for every mitigation in the Event Tree. Calculating the probabilities is already explained in previous paragraphs (10.4 and 10.5), depending on the type of ETA. In reality for different consequences, different combinations of "ideal" and "real life" ETA should be used. I cannot always classify

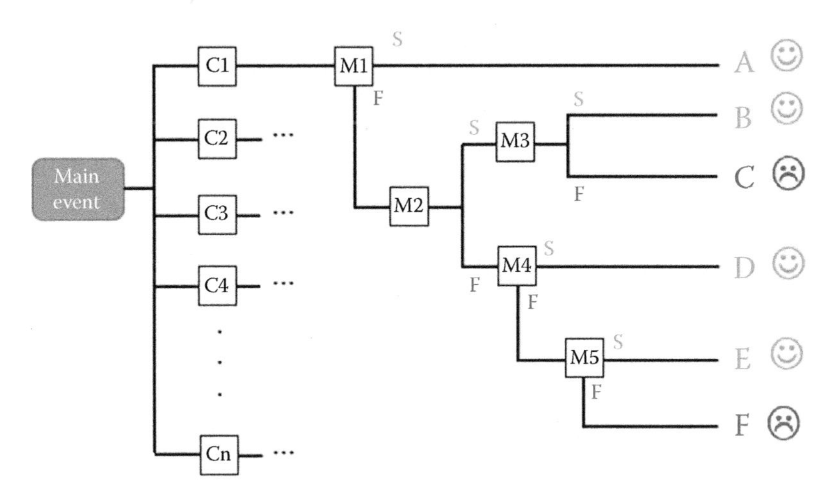

FIGURE 10.4
Combinations of "ideal" and "real" Event Tree Analysis.

the ETA as "ideal" or "real," so most of the consequences can be handled by a combination of these two (Figure 10.4). The important point is not to confuse you during the calculations of the probabilities for different ETA.

Figure 10.4 shows that there are three "ideal" mitigations (M1 and M4) and, in addition, there is M2 as "real life" mitigation. Mitigations M3 and M5 are our final choices, so they are not classified.

From Figure 10.4, I can derive these probabilities:

$$P(A) = P_1(S)$$

$$P(B) = P_1(F) \bullet P_2(S) \bullet P_{3|2S}(S)$$

$$P(C) = P_1(F) \bullet P_2(S) \bullet P_{3|2S}(F)$$

$$P(D) = P_1(F) \bullet P_2(F) \bullet P_{4|2F}(S)$$

$$P(E) = P_1(F) \bullet P_2(F) \bullet P_{4|2F}(F) \bullet P_5(S)$$

$$P(F) = P_1(F) \bullet P_2(F) \bullet P_{4|2F}(F) \bullet P_5(F)$$

Please note that there is no need to use conditional probabilities, as I have already stated in a few places in this book. I used them here just as an example, to be more accurate.

In general, the number of mitigations used in the ETA depends on the Main Event and its scenarios. In such cases, the BM's user safety experience is very important. The environment, knowledge, and data where, when, and how the reaction of a certain Main Event in the past was conducted are also factors in deciding which mitigation measures should be used.

11

Weaknesses of Bowtie Methodology

11.1 Introduction

Is the Bowtie Methodology (BM) the ideal methodology for Risk Assessment? Of course not![*]

There are weaknesses, and every user needs to know them.

Following the procedure for BM (FTA and ETA), you can notice that there are many steps that are sources of uncertainty. But let us be honest: The BM is a mathematical expression of reality that is based on the modeling of a system, data used to build a system, and data to make calculations. So, I can roughly classify the uncertainty in two categories: internal and external.

The internal weaknesses are expressed by the execution of the BM and the external weaknesses by the modeling of the system, the data used for modeling, and the data used for calculations. Both categories are subjective and strongly dependent on the person who is executing BM.

Scientifically speaking, the BM has a good mathematical background and there is no problem with this. But, the building of the FT and ET and the data used in calculations can also be a source of the problems.

Is this so considerable to make you not want to use the BM?

I do not think so.

Let me tell you a story to prove my point.

There is a well-known paradox about watches. You have two watches: One is not working at all and the other is 5 min late, but very precise. So, the question is: Which of these two watches is more accurate?

The watch which is late 5 min will never show the accurate time: It will always be 5 min late. But, the first watch will show the accurate time twice per day. So, by this context it is clear that watch which is not working is more accurate. But if I change the context, the result will change also. The main point here is the watch which is late 5 min and the question is: Knowing that it is late 5 min, will it help me to calculate the time?

Of course, it will!!! Knowing the error, I can produce accurate measurements!

[*] Do I know better methodology? Of course not! By in my humble opinion, this is the best methodology!!!!

So, the watch which is late 5 minutes is not accurate, but by using it I am able to calculate accurate time: Knowing the error I am able to achieve accuracy and precision in my measurements! And this is same with the BM: Knowing the places for uncertainties of the BM and their values, I can consider them in calculations. This will help me to improve the integrity of the overall methodology.

11.2 Modeling BM

I can state that I am using FTA and ETA as methods integrated in BM to "measure the reality" expressed by two aspects of a particular event: How the event happened and how will the consequences develop after that event has happened? Please note that I have used the words "measure" and "reality" in the previous sentence with "italic."

I did this for the "reality" because using FTA and ETA I am actually transforming reality into a model that I use to help myself understand reality. I did it for the word "measure" too because by using the BM I am not just explaining the reality connected with a particular event but I am also quantifying (measuring) this event with the appropriate calculations. In the calculations, I use particular numbers that express particular probabilities for the causes of this event. But, the point is that to calculate these probabilities, I build a Fault Tree (FT) and Event Tree (ET). The million-dollar quest here is: Would everyone build the same FTs and ETs?

Of course not!

The reason for that to build an FT and ET I need a particular knowledge about the system. I need to define the system using this gathered knowledge, which means setting internal and external borders. With other words: I need to build a model of my system that will represent "my reality"!

So, by the definition of the word model, it is my perception (my understanding!) of "reality" considering the system. Whichever system is modeled, the use of the BM will always depend on the understanding, knowledge, skills, experience, and creativity of the person who is executing it. And, this person is always a human full of subjectivism generated by its social history, education, experience, and position in society (and in the company).

There is no perfect model that means that there is always a gap between the model and the reality. This gap can be made smaller by including more people in the building of the model and that is something that I have explained in some of the paragraphs in the book. Looking for (and considering!) others' opinion will average extremes and help build a more realistic picture about the system. In the same time, it will (maybe) make some compromises that can endanger or improve the integrity of BM. So, there is need for a person who can take responsibility for the decision-makings (has a last word!) and the others are just there to provide information.

Whatever the cooperation of the other persons is, the person who is building the model of the system is coping with the responsibility of the integrity of the BM. Missing some of the specification of the system will make the modeling unfit for reality and undermine the results of the calculations whatever the FT and ET are.

11.3 Integrity of Data Used

Assuming that the modeling is OK, there is another important thing that affects the calculations. The integrity of the data used for probabilities is strongly correlated to the integrity (accuracy and precision) of the BM. The importance of building a realistic FT and/or ET is nothing compared to the integrity of data used inside. The accuracy of the calculations is connected with both the FT and ET and with the data used inside. It does not matter how well I do the modeling, wrong data will produce wrong results. It is similar with computers: The best hardware and best software (embedded into computers) will still produce wrong results if the data used for the calculations is wrong.

In the first few chapters in this book, I mentioned that I need more data to have more accurate calculations for the probabilities. As I mentioned there: Infinity is the best, but it is impossible to have an infinite source of proper data!!! So, I need to use a moderate number of data, which is actually another name for the data available.

But, be careful when using the Internet.

I read a very interesting story about data gathered from Internet. There was a project conducted by Harvard University professor Gary King for the prediction of unemployment by analyzing the use of Internet browsers. The project was based on the number of searches containing words like jobs, opportunities, and so on during the job search from unemployment persons. It is clear that if the unemployment is high, more people will use the Internet to search for jobs. Counting the daily use of these words will give us correlation to the employment rate. But, during the execution of the project, an incredible increase of data (searches for these words) was noticed and investigations showed that this was because of the recent death of Steve Jobs. People were actually searching the Internet for more information about his death (looking for Jobs) and the project had registered it as if the people were looking for jobs (working opportunities).

Anyway, to execute the BM, I need two types of data. The first type is data about the system and his operation, which will help me to build a model of the system. The second type of data is the data necessary to calculate the probabilities of different events which are contributing to the Main Event.

The data for building the model of the system should be found by the equipment manufacturers, manuals for operational procedures, and humans who are involved in the operation of the system. These are mostly internal sources of data and contribute a lot when building the model of a system.

The data needed for probability calculations are a problem due to two things.

First thing is that the data needed for the use of the BM in quality areas are not very available. The reason for this is that every company is treating their methods for achieving quality as their own company's secret. So, these data are available only for internal use. In these areas, where there is a lack of data, people will use any kind of data that they find, which is not very wise if you do not take care of the context.

In the safety area, using the BM is more rewarding due to the regulatory requirements to spread safety information all around the world. There are plenty of international organizations that are used as regulatory bodies for risky industries and they must gather data about the incidents or accidents from every investigation. Almost all of them are publishing annual reports that can be used as statistics for the probability calculations. Anyway, the main point is not to lose the context of these data. And let me say this again, gathering the data and their analysis is a responsibility of the person who is conducting the BM. He must be able to recognize the real context and value of the data used in the calculations.

Second thing is that the data on how humans react during an incident and accident have low integrity. These data are dependent on the company's organization, company's dedication to safety, the shape and integrity of the company's SMS, the safety training of the employees, the social and educational status of the company and its employees, and so on. There are many variables with different variability, so it is never easy to assume how the humans will react in case of a Main Event.

I strongly recommend to all of the engineers and managers who would like to use the BM to build their own (local) bank of data regarding the probabilities of the different events happening in their systems or in the similar systems based in their vicinity. Building these local data bank should be improved by the data gathered from other sources in the particular industry. As an example, I can mention that in aviation, plenty of organizations (ICAO, IATA, FAA, Eurocontrol, EASA, etc.) gather information regarding incidents and accidents and are regularly publishing particular reports with analysis at the end of year. All these reports are public, so you can find them on the Internet.

In addition, all manufacturers of equipment provide such data (mostly as reliability data), so do not hesitate to contact them! But, be careful! Very often, the manufacturers (for commercial purposes) "adjust" these data with the intention to "improve" the reputation of their products. So, be critical about the data taken from manufacturers. Record all problems (faults and failures) of the systems in your company and try to analyze the statistics behind them.

But, again, be careful.

A simple example: ICAO* publishes reports of incidents and accidents connected with air traffic through volcanic ash during volcanic eruptions. These data have nothing to do with the aviation hazard in Macedonia† because there is no volcano nearby and there is no record that there was in the last 2000 years. So, these data are completely irrelevant for the aviation in my country. Incidents and accidents caused by tropical cyclones are also not of interest for my country, so I do not need to list these situations as hazards.

The same thing applies for doctors. They must know the medical history of their patients to establish a proper diagnosis. So, I strongly believe that a good quality or safety manager must also have his local data bank with information necessary for the quality and safety systems used in his company or industry. The main focus should be on gathering data about the real situation in his company and these data must be used to follow and compare with the data about the situations in other companies from the same industry. These comparisons give you a good picture of your system (Is it OK or is it not?).

Another thing that you need to have in mind is: It is not enough to have a huge amount of data! You need to take care not to misinterpret these data! This is very easy if you miss the context and it can lead to a bigger problem where you begin to trust the data and you unintentionally make wrong decisions thinking that everything is OK.

11.4 Global Warming, Decision-Making, and BM

I would like to use an example here to emphasize a point related to the execution of the BM. This example is the global warming and all activities connected with this phenomenon.

Global warming is a term that was introduced in the late 1990s to describe the gradual increase in temperatures of air and oceans globally. The data gathered through the years, starting from the end of the nineteenth century, show that the air temperature is increasing. If this continues, there is fear that it will have devastating consequences on our lives and on Earth's ecosystem.

The main point stated by scientists is that all natural happenings at our planet can actually decrease the air temperatures. In the past, the Earth has experienced a few, so called, Ice Ages, where most of the Earth (both hemispheres) was covered in snow and ice. During these situations,

* International Civil Aviation Organization.

† Country where I was born is situated in south Europe in the middle of the Balkan Peninsula.

the polar ice caps were extended to the areas that are today moderate by climate and there were glaciers in the places between the mountains where today we enjoy moderate temperatures. These situations, on longer or shorter terms, were mostly caused by earthquakes and volcanic eruptions.

The last such short-term Ice Age happened in 1816 and it was caused by the huge volcanic eruption of Mount Tambora,* which happened on April 5 a year earlier in Dutch East Indies (Indonesia). This huge eruption sent huge amounts of gases into the atmosphere including sulfur dioxide (SO_2). Sulfur dioxide, in such quantity, caused a global change of temperature. As a result of this eruption, the global temperatures dropped and in the northern hemisphere there was no summer at all.† It caused a decrease in both food production and famine. Different sources assume that between 50,000 and 100,000 people died as a consequence of this "Ice Age."

Nevertheless, global warming is full of controversy. There are many people who disagree with it. Most of them state that the increase in temperature is evident, but it is within limits established by statistics (in the limits of standard deviation!). Some of them say that we cannot trust the measurements in the past due to inaccurate measurements, methods, and instruments that were not calibrated. Supporters of conspiracy theories stated that this is a hoax intended to colonize the Third World countries and make them stop using the CO_2 producing fuels.

The main point is that data in the last 50–60 years have shown a large increase in global temperature of the air and oceans and this is correlated to the increment of the CO_2 (carbon dioxide) emissions and the concentration of CO_2 in the atmosphere. Carbon dioxide is known as a "greenhouse gas," which is not allowing the surface of the Earth to be cooled naturally by radiating Earth's thermal energy into space.

I will not judge what is true in this case, but I know that some kind of a decision must be made. The point is, if I stick to the fact that there is an increase in air and ocean temperature and this trend continues, it will produce catastrophic changes in Earth's flora and fauna and badly affect our lives. So, even though there are two views (true warming or natural variations), a decision must be made based on the available data.‡

And this is true for everyday situations too: Sometimes, decisions must be made based on the available data, even though I cannot always trust (or suspect) the integrity of the data. The same things happen with the BM: Whatever probability I am calculating using the FTA or ETA, I need to make a decision, even though I may have doubts about the data used. So, by its

* This eruption is the largest known eruption on Earth. Before the eruption, Mount Tambora was around 4,300 m high and after the eruption, it was approximately 2,850 meters high (2/3 from its previous height).

† The year 1816 is known as the "year without summer" in the history of meteorology.

‡ So, on the basis of these data, I think that particular measures for limiting CO_2 emissions must take place!

ontology, it can happen that the BM cannot provide enough data for clear decision-making, but either way, you must make a decision. In a situation like this, it is wise to be more conservative (which brings you on safe side!), but it could be expensive. Using ALARP (paragraph 1.7 in this book) is a good practice to help you make decisions.

11.5 Weaknesses of FTA

Even though the FTA is an excellent method for analysis based on mathematics, it is not perfect.

The FTA is not very appropriate for dynamic systems. Dynamic systems are systems where some of the systems parameters are changing with time. These systems evolve in time by some particular rule (or without rules). Let us say that the safety FTA for building a house will change as the house progresses from its foundations to its roof. Similarly, the FTA for infectious diseases will change as the number of people infected changes. To be honest, you can still use it because there is no scientific reason not to. But remember that it may become too complex and this is not good because complexity produces chances for errors, which will diminish FTA's integrity.

The second weakness of the FTA is the fact that it deals with events that are very rare. This means that their probabilities of happening are very low and their meaning is not always clear. So, having in mind that decision-making is done based on these low values, it is better to deal with relative values. Let me explain what "relative values" mean in this situation. Let us say that the FTA showed me that the probability of having a certain Main Event is 2×10^{-6}. It means that if I try 1 million times I can have this event happening 2 times. But what does this really mean to me? The answer is: Almost nothing! But, if I implement some additional measures to deal with this and I do the FTA again, the result will give me valuable information. If the new FTA (after the additional mitigation measures) shows that the probability to have that Main Event now is 1×10^{-6} then I have achieved a result two times better than the previous one. With this result, the Main Event is happening two times rarer, which is good presentation and good achievement.

Another weakness is the fact the FTA deals with two situations only (failures and successes). If I am using it for systems where there is a third (or more) outcome(s), the situation will become very complex especially with the probabilities and dependencies (or independencies) of the events. Engineers who are dealing with digital electronics know that using the FTA for circuits where the SR flip-flop is used, there are four possible outputs. But, two of them, one being a situation with input $S = 0$ and $R = 0$, will produce an output that is dependent on the previous state of the flip-flop and the other situation with input $S = 1$ and $R = 1$ will produce an unpredictable output (0 or 1).

This means that the FTA is useless for circuits with SR flip-flops. It is similar with circuits using the JK and T flip-flops, where there is a dependency on the previous outputs for particular combinations of inputs. Anyway, even in such cases, the FTA is still applicable, but the fault tree should take care for these characteristics and this can make the analysis very complex.

Some authors mention the incapability of the FTA to deal with common mode failures and this is something which I cannot support. By definition, these are failures that are dependent on each other and they cannot always be identified because they are usually latent conditions inside the system. It is still possible to use the FTA in cases like this if you identify these common mode failures, but some complexity regarding the fault tree and the calculations might arise. The main problem here is that the common mode failures can produce additional accidents due to the incapability to determine the real connection between them. Considerable working experience can help in those cases.

11.6 Weaknesses of ETA

If you consult the literature for ETA, you will find different weaknesses depending from one to author another.

The first one is that there is no standard for the graphical representation of the event tree. I do not think that this affects the integrity of the ETA. So, feel free to use whatever you like for presentations, but take care to build a realistic scenario for the development of the consequences and to use data with high integrity for the probability calculations.

Some authors mention that the ETA is not holistic because you can only study one consequence of Main Event in each ETA. This is true, but if you produce an ETA for each consequence, it will make it holistic. This is also not an issue because with the ETA I am more concerned about the results of the work than the volume of the work. Using today's computers, calculations can be done very fast, so this is not a big issue.

There is also a danger of not understanding the dependencies between the components of the system included into the Main Event, but this is a thing that is applicable to all methods. Anyway, this can trigger another problem that is connected with common cause faults or failures during calculations. Of course, this is something that needs to be considered during the creations of the scenarios. The important thing is not to miss some of the components or events in building these scenarios or neglect some of the dependencies between them.

Sometimes there are few types of mitigations that can be implemented on one Main Event or during the development of scenarios, which is actually an opportunity for you to choose the one you would implement first. In this case, it is wise to do an ETA for each of mitigations chosen as if it was the first in order. This will make it easier to determine which one is more effective and efficient by looking at the calculated probabilities.

The ETA may sometimes be complex, but simplifying it is not always the solution. Presenting complex scenarios with a simple model just for the sake of simplicity is a wrong attitude. If you need simple methods, use simple systems!

Similarly to the FTA, the ETA can experience problems when representing scenarios for systems that are dynamic (evolving in time). But the problem here is that the Main Event will evolve and the ETA should follow this evolution. This will produce different scenarios for different states of the Main Event and producing an ETA for such situations is very ungrateful.

However, in my humble opinion, the biggest problem with the ETA comes from its fundamentals. ETA deals with scenarios of how the possible consequences will develop. The main point is the word "consequences"!

There are two types of the consequences: direct and indirect. Direct consequences are those that affect our situation immediately: injuries, deaths, assets damaged or destroyed, shock in the humans, change or incapability to maintain or sustain working routine, and so on. After the incident or accident happens, the indirect consequences are hidden. They show up later, when I am trying to restore normal operations after the first mitigation measures are implemented. Indirect consequences show up as costs to deal with injuries, deaths, and/or resuming normal operations. Here we can also add losses due to stop of production, losing moral in the company, and losing confidence from customers.

Direct consequences are usually the subject of FTA and ETA. If they are executed in advance, it helps very much in eliminating and/or mitigating these consequences. Usually, measures triggered by ETA are some kind of backup plans that are immediately executed. However, indirect consequences are usually not a part of ETA and that is the reason why they may appear later as more critical. It is wise to deal with the indirect consequences in advance, but, unfortunately, no one is thinking about them when everything is OK. Whatever the warnings about incidents and accidents, they are usually neglected by most of the managers due to economic orientation of the companies. The main question is as follows: What is the cost of every incident and accident?

In the area of aviation, such a calculation was made in the USA in 2003. This was a time when the implementation of a Safety Management System in the aviation subjects worldwide became a regulatory obligation. Airlines in the USA were concerned about the costs of the implementation and they complained about it. In the effort to respond to these complains, the White House established Commercial Aviation Safety Team, which was tasked to investigate the costs. In the Report* provided by the Team to the US Congress in 2004,

* Report on Commercial Aviation Safety Team; White House; USA; 2004.

the possible benefits from implementing SMS in aviation were presented. One of the findings from the report was:

> The decrease of 73% in safety risks will save the airlines 620 million US dollars every year. Every safety incident (compared by the number of flights) cost aviation subjects 76 US dollars per flight. By implementation of only 46 recommended safety improvements, this cost decreased to 56 US dollars per flight.

The facts in the previous citation are telling us enough of what to do, but anyway, this was neglected many times in the past. So, I will not comment on it anymore.

12

How to Improve Safety with Bowtie Methodology

12.1 Introduction

The Bowtie Methodology (BM) is used for gathering knowledge about risky events and how these events transform into consequences. With other words, it is a methodology that we use to diagnose the type and nature of a certain "disease" with the intention to prescribe the best "therapy" available. So, the BM establishes the "diagnostics," but we need to decide what the "therapy" is, and I think it is a beautiful tool for such a thing! To be clearer: I would like to point that BM is helping us to find proper place where to put elimination or mitigation measures, but which one, we need to decide.

There are many ways to improve the safety of systems when you know the real state that they are in. As I already said in paragraph 1.1 of this book, the BM consists of two methods: FTA (Pre-Event Analysis) and ETA (Post-Event Analysis).

Measures for improving safety can be implemented in both of these methods. It is wise to document all of these improving activities with the intention to follow their effectiveness and efficiency.

12.2 Preventive and Corrective Measures

In safety (and quality also) science, there are two types of measures that help to deal with incidents and accidents: preventive and corrective.

Preventive measures are undertaken with the intention to prevent certain "bad things" from happening or to eliminate or mitigate the consequences in advance. Usually, these are things that do not contribute to the normal functioning of the system. But they are important because they are installed with the intention to protect the system, humans, and the environment from the abnormal functioning of the system and to mitigate consequences (if such abnormal functioning happens). A good example for this is the airbags in the cars.

They do not influence the characteristics of the car but protect the driver and passengers in case of a crash.

Corrective measures are implemented in the system if the safety or quality analysis shows deficiencies in normal operations. These deficiencies can endanger the safety or quality in normal operations, so corrective measures are implemented with the intention to correct deficiencies and provide better operating of the system. If I notice that something is wrong with my car, fixing the problem in a workshop is a corrective measure.

Preventive and corrective measures can be implemented into both FTA and ETA. Which one and where you will use it depends on the experience of the operator. In general, they may apply to all three constituents of the system: humans, equipment, and procedures. The best solution is to deal with equipment. Dealing with procedures is maybe the cheapest, but it can work properly only if the safety management system is good and there is a culture for safety support in the organization. Humans are the weakest link in this chain, but their behavior is strongly dependent on the type of organization and managers inside.

At the end of paragraph 4.4 (Canonical or Standardized forms of Boolean functions), I explained that the BM in the area of FTA (through minterms and maxterms) explains the scenarios of how "bad things" happen. So, I can use this knowledge to change something in the diagram with the intention to eliminate or mitigate the risk. It means that I am putting there some kind of a measure or activity (usually called barrier). Implementing these additional measures in the FTA (Pre-Event Analysis) means that I am undertaking activities that will help reduce the risk. These measures can be implemented in different levels (iterations) of the FTA (as shown in Figure 1.6), and they need to be checked regularly on how successful they are. The main point is that if I do that, I need to change the FTA diagram and recalculate the whole FTA following the newest diagram with these additional measures.

Actually, these measures in the FTA will decrease the probability (frequency) of the Main Event happening and they can also (maybe) decrease the consequences if the Main Event happens.

Let us give one simple example for this: There were data that on a particular part of the road outside the city, incidents and accidents happen very often. I did an FTA and noticed that the main causes for this is the high speed of the cars and the bad perception from the drivers regarding the opposite traffic due to the curvature of the road. There was a sign for a speed limit of 60 km/h and there were also buildings that did not allow the drivers to see the traffic from the opposite direction. So, limiting the speed to 40 km/h and putting a traffic mirror for the drivers to be able to see what is behind the buildings will reduce the probability of incidents or accidents (will reduce the risk!) and at the same time reduce the consequences of possible crashes due to the lower speed (40 km/h instead of 60 km/h). So, I intervened in Pre-Event analysis (put traffic mirror and 40 km/h limit of speed) and in doing so (by producing lower consequences) influenced the Post-Event analysis.

The second possibility is to deal with Post-Event analysis (ETA) and eliminate and/or mitigate the consequences. Dealing with risks in the ETA can be good only if the elimination and mitigation activities implemented do not create new hazards because this means that they must be analyzed again in the Pre-Event analysis (FTA). To be sure that no other hazards are created, I need to check again the FTA (with these activities inside) and the probability calculations.

The first thing that needs to be done by the ETA is to provide activities for the containment of particular consequences. In firefighting, this is known as suppressing the fire (limiting the scope of fire, before fighting it). The second thing is to check if the Main Event can trigger some other (secondary) events that can be also (more or less) dangerous. The ETA should first explain how to stop these secondary events and after that produce scenarios explaining how they will progress. A simple example here is an outburst of fire in a chemical factory. The most important thing is to contain the fire at a safe distance from the chemicals, because if it spreads, it can cause an explosion or release poisonous gases in the air (atmosphere).

After that, the ETA should focus on dealing with the consequences of the Main Event and produce procedures on how to mitigate them.

12.3 Risk Control

To deal with the risk I need to take care about it through appropriate control. By control of the risk, I mean the activity of monitoring the situation and implementing particular adjustments (if necessary!) to stay in the limits of tolerances for the system. Risk as a risk, as big as it is, may not happen if I detect it in a timely manner and it cannot happen if I continuously monitor the situation and immediately apply particular adjustments. Keeping an eye on the road when you are driving your car is a typical example of monitoring. Of course, monitoring will trigger some actions with the intention to drive safely. So, monitoring the present situation, thinking in advance about the situations on the road, following the traffic rules, adjusting the driving on the environment, and not driving aggressively have kept many drivers far away from incidents and accidents.

Monitoring is a process in which the functioning of the system (equipment, operation, activity, process, etc.) is followed by a person or device. Monitoring may rise alarm or show the need to implement some automatic measures if the operation deviates from the normal operating. Anyway, if the monitoring is done by a person or by a device, there must be procedure on how to react in the case of abnormal situations. Even though there are some general measures that can be helpful in particular situations (executing particular procedure), there should be a specific procedure for each abnormal situation. Procedures implemented for abnormal situations take control of the situation in case the risk is close to being materialized.

There are plenty of control measures that can be implemented and they differ from situation to situation. In general, I can categorize them from the place where they are applied and from the time when they are applied. In addition, all of them can take control over hazards or risks.

Hazards and risks can be eliminated or substituted. Risks also can be mitigated by decreasing their severity and frequency (probability).

Elimination is always the best! For the BM, the elimination of risks should be in the FTA. Not applying for a hazardous job, not using a hazardous tool, process, machine, or material is the best way to get rid of incidents and accidents.

Why in the FTA? Because the FTA is at its best if executed during the design phase! Do not forget, ETA is applied after Main Event happens.

However, there is something more important with all these control measures and that is how they change the FTA and ETA! Every control measure that you implement in any part of the system under consideration will create new hazards. If you put a device that provides an automation of a process, this device is part of equipment that has its own reliability too. And this reliability must be calculated in the FTA and ETA. So, whenever you introduce a new measure with the intention to decrease the probability and mitigate some consequences, an update for the changes in the FTA or ETA must be done and the probabilities must be calculated again.

12.4 Control by Feedback Circuit

Feedback is a way to control the electronic and mechanical devices by implementing a close loop* when the output signal is compared with the standard and if there is difference, the circuit will generate information. This information in the form of corrective signal is sent to the input (Figure 12.1). There is no better way to control a system inside the tolerances for normal operation than feedback. Please note that in Figure 12.1, the control of a normal functioning system (success) is explained.

This is part of the automation applied for the control of normal operations and it is an extremely effective and efficient way to control the situation. It is effective because the delay in the reaction is very small and it is efficient because small changes in the input result in big changes in the output.

* In literature about dealing with control systems, ***Feedback*** is a synonym for ***closed loop circuit***. There is one more method for control—the so-called ***open loop*** and it can also be found as ***Feedforward***. While feedback is extensively used, Feedforward is not very popular.

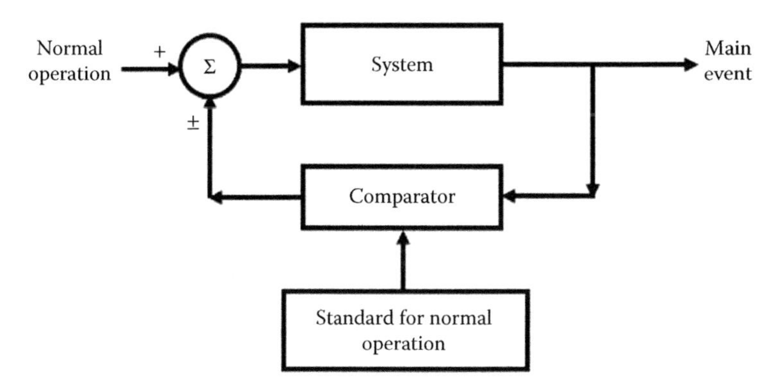

FIGURE 12.1
Feedback circuit for control.

In this case, the Main Event (normal operation) is controlled by a circuit (Comparator) that has two inputs and one output. The output of the Comparator is connected to the summation circuit, which is used to increase or decrease the input (which produces the change in the output). Two inputs are connected to the Comparator: The signal from the output of the System (the normal operation!) and the signal presenting the Standard for normal operations. They are compared and if the difference between them is bigger than the specified tolerances for the normal operations, then a corrective signal is sent to the summation circuit (presented by S), which changes the value of the input. This is usually done with the intention to keep the normal operation of the system intact, especially in situations when too high input can cause saturation of the output. If a failure happens, then the Comparator will shoot down the system and contain the consequences. Some of the feedback control circuits have the option to reset the system during an operation, but it depends on the system.

A good example of a feedback as a close loop control system is the thermostat of iron. Before ironing, I adjust the knob to the requested temperature for the clothes that I want to iron. When the temperature is achieved, the thermostat will switch off the iron. When the temperature falls down to the temperature determined by the hysteresis of thermostat, the iron will be switched on again and the process will be repeated again and again keeping the temperature inside the tolerances of the system (ironing).

There are two types of feedback: positive and negative. Positive feedback is presented with a plus (+) in the input of the summation circuit (S) in Figure 12.1, which means that the Comparator is producing a signal that is telling the system that it is functioning okay (by the standards). Negative feedback is presented with a minus (–) in the input of the summation circuit (S), which means that the Comparator is producing a signal that is extracted from the signal for normal operation. Negative feedback is mostly used in automation electronic circuits for control. Positive feedback is used almost everywhere, but not as often as the negative feedback.

This is an excellent way to deal with successes (providing normal operation), but it is also applicable for failures (as containment for abnormal operation) of the systems. It is applicable to both FTA and to ETA, but you need to be careful not to forget to apply the reliability of the feedback circuit into the FTA and ETA calculations.

12.5 Measures for Risk Elimination and Mitigation

If I am not able to eliminate the problem, then I should try to substitute it. This means that if something is hazardous (machine, tool, or material), I need to find a substitution for it. For example, a hazardous material (asbestos, mercury, etc.) can be replaced by a less hazardous one. But, there is another aspect of substitutions and that is replacing humans with machines in areas where humans are at risk.

Elimination and substitution can be done by both engineering and management methods. It is important to re-evaluate the FTA and ETA after some of the measures are implemented, because every measure introduces a change in the system that can introduce new hazards.

Let us talk about some of them.

12.5.1 Redesign

Redesigning equipment, tools, or processes is connected with the FTA executed during the design phase of the product or operation at hand. Executing the FTA in this phase will trigger a change in the design and it will save on costs for later redesigning. Problems are not always noticed during the design phase and this mostly happens when a company is trying to launch a new product or operation when there is a deadline. In rush to keep up with the deadline, some tests are not conducted and the real picture about the product or the operation is not perfectly clear, so the change in the design can be triggered even later, after the product is sold to the customers.

For example, from 2005 to 2014, Opel, a German car manufacturer, had a problem with plenty of its models sold in the UK and Ireland due to fire in their cars. Fires were triggered by the improper design of the fan and the bad components in car's cooling system. The hazardous system was redesigned and the problem eliminated! After that change, no more fires were reported! It cost Opel a lot of money, but the problem was solved.

Let us talk about another interesting story about redesigning.

At the beginning of the second half of the last century, the first machines for cutting books in proper shape after printing were made with one switch for command to push the sharp and heavy knife electronically down for cutting. This resulted in a lot of workers losing their fingers and palms. Investigations showed that the workers were using one hand to adjust the samples and another hand to press the switch. In a hurry to make more products (earn more money!), they made mistakes that costed them their fingers and palms. So, designers decided to put two switches in series to make the workers use both hands to press the switches. The injuries decreased, but they did not disappear. Further investigation showed that the switches were too close to each other, so again the workers in a hurry for more products used one hand to adjust the samples and different fingers from the other hand to press the switches. After another redesign, the switches were put at 1-m distance, so the workers would have to use both hands to press them. The injuries were eliminated.

12.5.2 Isolation

Sometimes I should try to isolate, contain, or keep away the hazard from the operators or the environment. For example, the PPE (Personal Protective Equipment) as part of OHS (Occupational Health and Safety) regulation intends to isolate the workers from some of the hazards. Going further, an insulated and air-conditioned control room can protect radiologist from the X-rays in the hospitals or protect the operators in the nuclear power plants from radiation. Biohazards in medical laboratories are prevented by the isolated working places that allow the personnel to work without endangering their lives. Hazards are still present, but they are isolated, meaning that they cannot do any damage to the humans and to the environment. Isolation is applicable to both FTA and ETA. Even if the Main Event happens, the consequences can be decreased by measures of isolation (quarantine of the patients in the case of infectious diseases, etc.).

There is a beautiful example of isolation in aviation. As I mentioned before, all equipment and humans in aviation are doubled. There are two pieces of CNS equipment in the cockpit, there are two persons inside the cockpit (pilot and co-pilot) and there are two engines in every commercial aircraft. There is a rule of a thumb that the regular checks and the maintenance of the aircraft engines must be done by two independent teams (one team per engine). The reason for this is that if there is some systematic mistake in the engine made by one of the teams, then this mistake will not be done on the second engine. So, such a mistake in one engine is isolated only to this engine and it will not affect others.*

12.5.3 Automation

People, in general, tend to automate or mechanize the processes and this is very useful when applied to risky processes. Actually, this is part of the

* This is known as the isolation of Common Mode failures.

substitution: The hazard is still present, but the humans are replaced with automatic machines (equipment), meaning that the humans do not experience any harm at all. Because of the fast technological development, almost all processes are automated by computers. A great example is the computer-controlled robots that handle the work of nuclear reactors and the production of oil in refineries. This is a funny situation because the automation dealing with one or few hazards also creates new hazards. As I said before, the use of a robot means that I need to calculate his reliability too. I also need to check the changes in the FTA and ETA. An example for automation is the feedback circuit for control (paragraph 12.5, Control by feedback), which is very often used in electronic equipment.

12.5.4 Barriers

They are extensively used in safety in both areas: Pre-Event and Post-Event activities. Barriers can be used to stop the causes for hazards and risks, but they can also be used to stop the spreading of the consequences of incidents and accidents. There is even a so-called *Swiss cheese* model,* which explains how accidents happen even though the barriers are in place.

Particular barriers can block hazards before they happen or before they reach the humans. For an example, special protective glasses can prevent eye injuries from UV welding arc radiation, just as a special UV cream can protect the skin from the sun, and so on. In addition, it can stop the consequences after the event has happened: Proper clothes can protect firefighters from fires, divers from hypothermia or high pressure, cosmonauts from vacuum and lack of pressure in the space, and so on. Historically, barriers are the oldest safety measures that protect humans, assets, and environment.

There are three types of barriers: Passive, active, and procedural.†

Passive barriers are passive elements that are not using any kind of energy to protect something from some bad influence: protective glasses for the eyes during welding, walls intended for thermos protection during fires, and so on.

Active barriers on the other hand use some energy to protect the humans or assets: air bags in cars, emergency buttons on equipment, fire protection system in buildings, and so on.

Today, the procedural barriers are both the most important and the most neglected. The reason for that is the lack of awareness by managers that human and organizational errors are endangering lives. Procedures are used to connect humans and equipment (Operational procedures) or to establish a particular management system in the company (System procedures). Sometimes a simple change in procedure can make a huge difference in the probability of an accident. Procedural barriers are rules that included in procedures and if

* See paragraph 1.2 in this book!

† Different authors have different classifications, but I prefer this one, which is the simplest and totally fits the purposes of this book.

followed they stop "bad things" from happening. An example for this kind of barrier is the procedure of driving at crossroads where traffic lights are present. Procedural barriers are mostly dealing with "loss of control."

Even though the procedural barriers are important, I strongly recommend that you do not use them. The reason for that is that these barriers are executed by humans and, as I already said: Humans are not reliable. Whenever you have the chance to use a mechanical or electrical circuit to put up a barrier: Use it!

In nuclear industry, there is the requirement to have at least three barriers in the FTA, but medicine usually relies on only two (in the ETA): The capability of the doctor to diagnose the disease and prescribe good medicine that will cure the disease.

If you search the literature for barriers, you can notice that the quality of the barriers is measured by so-called, ***escalation factors***. Escalation factors are used extensively in the Bowtie method (petrochemical industry!) and they present factors whose values make the risk unsustainable. If these values are passed, they change the risk, making it intolerable.*

12.5.5 Absorption

The incidents and accidents (in practice) happen when the humans and assets cannot absorb the adverse energy created by malfunction of operations or natural disasters without harm or damage. If humans or assets could absorb these energies, then there would be no need to protect them from harm or damage. The seismic buildings in civil engineering are a good example for creating assets that can absorb all energy from earthquakes and protect the humans and assets from disasters. In addition, this is also used in car safety today by putting particular devices in the cars like seat belts, air bags, and so on. They have already proven themselves by absorbing adverse energy during car crashes and helping humans survive them.

There is particular discipline in safety science called Resilience engineering† that takes care of producing systems (equipment, processes, operations, etc.) that will be tough enough to absorb known hazards and elastic enough to deal with unknown ones. Of course, this should be applied during design phase.

12.5.6 Dilution

This is a chemical term that can also be applied in Risk Management especially in the chemical industry, medicine, and pharmacy. There really are

* Following this definition, I may say that tolerances in normal operations are escalation factors for normal operations.

† Resilience engineering is based on the ideas of Japanese quality guru Dr. Taguchi. He was speaking about "robust systems" that are ready to absorb all adverse energies.

some hazards in industry that can be diluted (or dissipated). For example, dangerous acids spilled on human skin can be diluted by water, ventilation systems can dilute toxic gasses with air before they reach the outside world, and so on. The Union Carbide factory in Bhopal (India) had implemented such measures to dilute methyl isocyanate (MCI), but it was not effective. The water pressure in the hoses was not strong enough to reach the "chimney," where the gas was freed into the atmosphere. It was the last barrier of protection* and its failure resulted in an accident (1984) that killed somewhat between 2,000 and 3,000 citizens.

Dilution can be applied both in FTA and in ETA.

12.5.7 Oversight

Oversight† activities may be internal or external. Internal activities are actually activities of overseeing the normal operations and differ from monitoring by nature. Monitoring is a continuous operation and oversight is scheduled. It can be formal as quality audit (requested by ISO 9001) or it can be arbitrary when managers would like to check some activity. Regulatory oversight, by definition, is external oversight and is scheduled (periodical by regulation). This is oversight where regulatory bodies (especially in risky industries) are looking for proofs that company is providing save operations.

Oversight activities are excellent measure for convincing company, regulators, and public that risk is managed properly and normal operations produce no risk for humans, assets, or environment.

The main benefit of these oversight activities is that they can find fundamental organizational errors that are hidden and may result in accidents. The internal oversight is not so good compared with external oversights. Reason for that is the subjectivity that can be present if a company has organizational issues. This is oversight that is executed by method "oversee me by me," so it cannot always provide objectivity necessary to get the real picture of situation.

Regulatory oversight shows status of organization and their safety culture and bring objectivity. The main problems with regulatory oversight could be when auditors are following the bureaucracy instead looking for the reality of situation.

Oversight activities (by themselves) will not change the FTA or ETA, but their results may trigger activities for the elimination or mitigation of some organizational errors, which is actually changing of the operation of the system. These changes must affect FTA and ETA.

---

* Actually, there were three barriers and all of them were not working. The water dilution of MCI was the last one.

† In this paragraph, I treat Oversight as Audit Activities (internal and external). It is different than continuous monitoring.

13

Future of BM

13.1 Introduction

As I said at the beginning of this book, the Bowtie Methodology (BM) is a holistic approach for analyzing safety because it contains two main things that complete safety: Cause analysis (Pre-Event analysis) and consequence analysis (Post-Event analysis). Cause analysis is done with the FTA and consequence analysis done with the ETA. The name (BM) is coming from the shape of these two analysis connected with the Main Event.

Theoretically, you could use any other method instead of the FTA and ETA for the Pre-Event and Post-Event analysis. You would have something with the same purpose, but it will not be the BM. The reason for that is that for the time being there is no better connection between two methods than the FTA and ETA, meaning that the BM is unique.

The FTA and ETA were first used almost 40–50 years ago and they proved to be very successful. Of course, they are not perfect, so scholars started to investigate their deficiencies. And when they did find weaknesses, they started to look for ways to improve them.

In this chapter, I will mention (only as information!) some of the improvements of the FTA and ETA that are somewhat popular at the moment. Their presence in industry is not very high for the time being, but this does not means that it will not change in the future.

13.2 Future of FTA

There are plenty of variations of the FTA with the intention to categorize it for different purposes. The most used are the Dynamic FTA and FTA combined with Markov Chain Analysis (known as FTA+MC). There are also new methods derived from the FTA that arose from its critical assessment. These are Hip HOPS, Component Fault Tree, State-Event Fault Trees, and Binary

Decision Diagrams (BDD). All these variations and new methods of the FTA are just efforts to simplify and improve the real FTA.

The point here is: If the FTA is fully applied to complex systems with wide external and internal boundaries, it becomes very complex, so classical FTA will not always provide clear answers.

Let us go briefly over them.

13.2.1 Hip HOPS

HiP HOPS stands for Hierarchically Performed Hazard Origin and Propagation Studies. You can see the diagram in Figure 13.1. The name comes from the fact that it determines where the hazards have originated and how they propagate in a complex system.

This is a method (presented with a particular software) that is an aggregation of a number of classical methods such as FTA, FFA,* and FMEA.† It works by integrating these methods in order to address some of the problems associated with the design of complex technological systems. The assessment of complex systems is done starting from the assessment of their functioning, registering the part where the fault or failure first happened, and finishing with the assessments of the components that have failed. It is a simple method using the simplified development of Fault Trees, but it can guarantee a better consistency of results than the FTA.

The application of the method starts by presenting the function of the complex system (FFA). After that it continues with the FMEA, which is used as

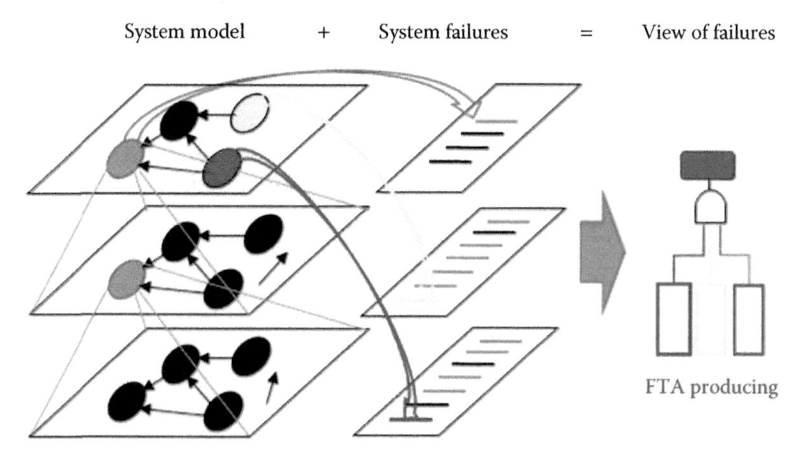

FIGURE 13.1
Hip HOPS structure and phases.

* FFA stands for Functional Fault Analysis.

† FMEA stands for Failure Mode and Effect Analysis.

an interface between the functions of the system and parts included in it. Eventually, the method finishes with a simplified FTA. Actually, this method connects the functional failures (determined by the FMEA) with the component faults (explained by the FTA), providing a clear picture of what failed and how: function or component.

13.2.2 Component Fault Tree

The Component Fault Tree (CFT) method is an extension of the FTA that is especially useful in embedded systems.* Embedded systems are part of the mechatronic area and these mechatronic systems are made of mechanical and electrical parts. By using the CFT to analyze the safety of these devices, I actually conduct a "damaged" FTA. I am saying "damaged" because I am not building a full system fault tree, instead, I am dealing only with components that are part of the failure scenario. When you have a complex system under analysis, the CRT can produce a less complex fault tree where only the components included in the failures are present. Either way, the CFT produces a good connection between the architecture of the system and its functionality.

The CFT is built of same elements as the FTA, but, in addition, it has input and output failure ports and internal failure events. A simple CFT for a fire extinguishing system using water is shown in Figure 13.2.

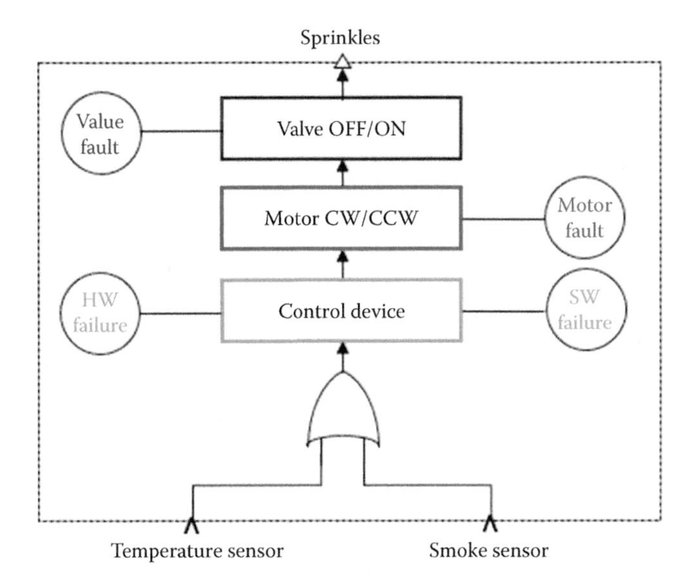

FIGURE 13.2
CFT for fire system with sprinkles.

* Embedded systems are processor-controlled devices made as part of some mechatronic device with particular function (usually to monitor and control the operation).

Input and output failure ports are presented by the open triangles that are used to show the possible points of fault propagations. Internal failure events are presented by circles and are used to present basic events in the FTA. So, the CFT is making fault trees for each output failure port and they can be analyzed as a function of the input failure ports and internal events.

13.2.3 State/Event Fault Tree

The State/Event Fault Tree (SEFT) is very popular when dealing with software. Considering that all complex systems today are powered by software, the SEFT can be applied to all complex systems. This method is making a difference between the software functioning (particular state of program execution) and the hardware faults (for this particular state of software functioning) through the notion of the events and states of equipment in the FTA. The events are sudden happenings that represent the failures of the system. The states are events that last some time and are an expression of the proper functioning of the system. SEFT is actually a continuation of the CFT and it is good for dynamic and complex circuits.

The SEFT helps to get a clear picture of dynamic systems, especially of their timings, statistical dependence, and sequence of the operations and events by using different gates for states and events. Let us say that I have two OR gates (one for states and one for events) that are connected by an AND gate. This is done with the intention to explain the overall contribution of the states in combination to the events in the functioning of the system. In addition, there are Historic-AND, Sequential-AND, and Priority-AND gates, which provide a better description of what is going on and when.

Figure 13.3 shows a SEFT diagram for two parallel processes that are controlled by a controller and a set/reset by R/S circuit if both of

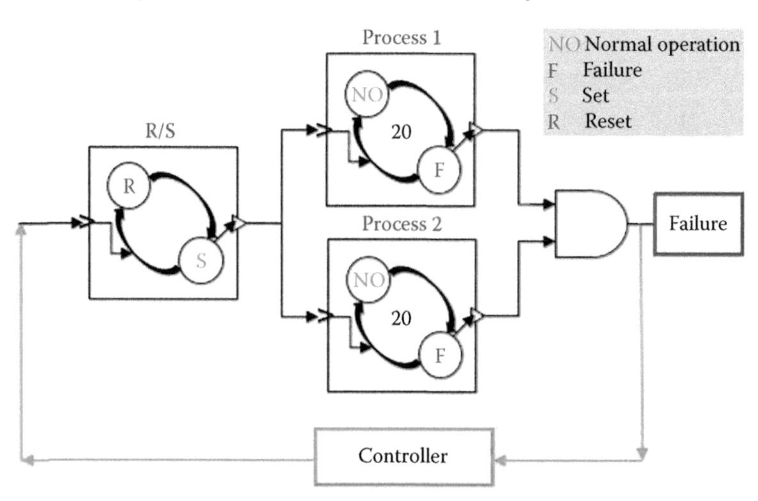

FIGURE 13.3
SEFT diagram for two parallel processes.

them fail. These processes have an MTBF equal to 20 h. There is also a Controller that senses faults and uses S/R circuit to set and reset the processes accordingly.

The SEFT is not so efficient compared to other FTAs and the reason for this is its complexity and the number of different notations used.

13.2.4 Binary Decision Diagram

Binary Decision Diagram (BDD) is a graphical representation of the Boolean functions, similar as the FTA. The difference is that with BDD I am building a BD tree, where there are no Boolean gates. Actually, the steps that I am following are triggered by the inputs.

The BD tree for AND and OR functions of two variables is given in Figure 13.4. You can notice that the violet 1 (level 1) is the first variable (so called "node") and the violet 2 (level 2) is the second variable (two nodes for two values of x_1). They are connected with lines, expressing the combinations between their values (dashed line for 0 and full line for 1), which I call "roads." Following the road expressed by the red arrows of the AND function (Figure 13.4), I get the value of the function F for $x_1 = 1$ and $x_2 = 1$ ($F = x_1 \cdot x_2 = 1 + 1 = 1$). The same thing is presented (on BD tree) with a blue line for the OR functions for $x_1 = 0$ and $x_2 = 1$ ($F = x_1 + x_2 = 0 + 1 = 1$). So, different roads express different situations.

As we have already seen, the BDDs are based on BD trees, and they are much simpler than the FT (fault trees in FTA). In BDD, I can omit some of the nodes and the lines. Looking at Figure 13.5, I can see that the AND and OR functions from Figure 13.4 are presented as BDD. In Figure 13.5a at the AND function, x_2 is omitted because whatever its value, the result is 0. In Figure 13.5b (the OR function), I have omitted the right x_2 because whatever its value, the result is 1.

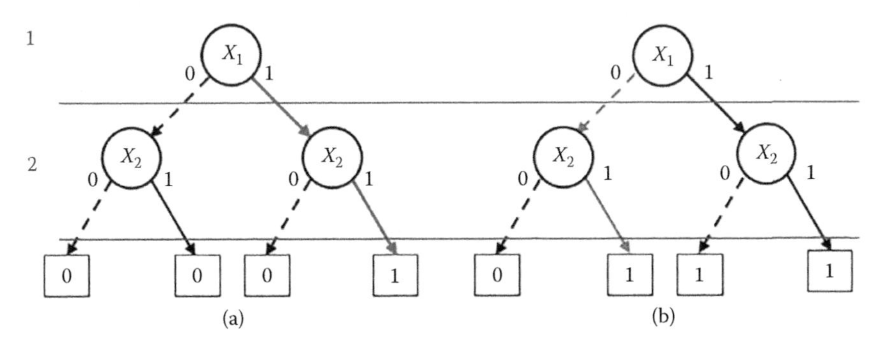

FIGURE 13.4
BD trees for AND (a) and OR (b) functions.

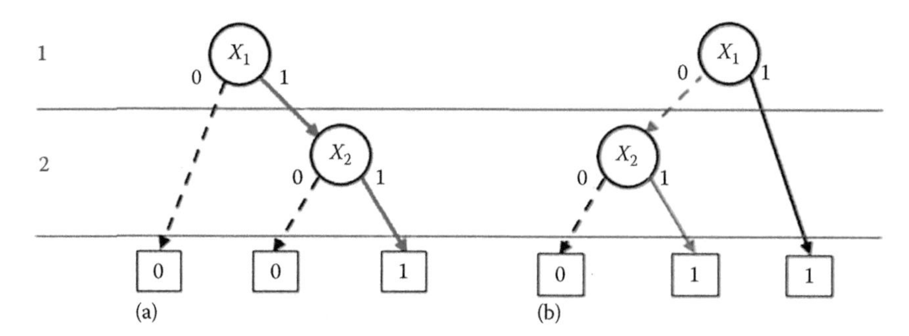

FIGURE 13.5
BD trees transformed into BDD for AND (a) and OR (b) functions.

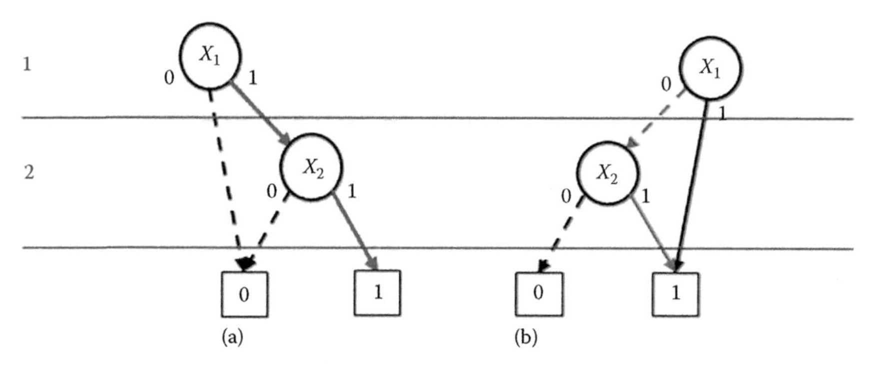

FIGURE 13.6
Simplified BDD for AND (a) and OR (b) functions.

I can go further with the simplification, omitting the values of 0 for the AND function and value of 1 for the OR function. Both functions in their simplified versions are presented in Figure 13.6.

BDD can sometimes turn out to be huge because it is increasing exponentially by the number of variables. Nevertheless, it can be very useful for complex systems. The size of the BDD is defined by the number of nodes (levels) and by the length of the longest computation path of the BDD. The longest computations for our example are the red and blue lines (AND and OR functions).

13.2.5 Boolean Logic Driven Markov Processes

As the name says, this is an FTA for the Markov processes. The Markov process is a stochastic* process with a finite number of possible outcomes or states. The output of every stage depends only on the output of the previous stage.

* Using simpler words, a stochastic process is a process where every next step (inside the process) cannot be determined with accuracy (it depends on some probability).

This is a method that has a diagram closely related to the fault tree. It was proposed in 2002. It provides few advantages: easiness with ambiguities, simple readability, and the possibility to take some of the dependences (that could be neglected during the building of the FT) into consideration.

The Boolean Logic Driven Markov Processes method uses arrows to present the transfer from a normal to a failure state and/or vice versa. This method is very complex and beyond the aim of this book. Even though I will not go into details, it was necessary to mention it here.

13.3 Future of ETA

Even though the ETA is used for cause or consequence analysis, it does not develop with the same pace as the FTA. At the beginning, the ETA was used for Pre-Event analysis in the nuclear industry and it can still find its purpose there, especially in the area of estimating consequences. In nuclear industry, the ETA has developed into a Containment Event Tree (CET) as a tool used for analyzing the development of the consequences in time.

13.3.1 Containment* Event Tree Analysis

The CET analysis deals with the containment of nuclear material after an incident has happened. It is used to estimate the range of the consequences of releasing the nuclear material, which are corresponding to the particular incident scenario. An important point with the CET analysis is that it is probabilistic and does not only take the outputs into account but also the time frames of each scenario (how it develops in time!). This means that the probabilities are important, but not as much as the time (time frame) needed to deal with the incident with the intention to contain the nuclear material in the reactor and not to allow it to endanger the environment outside.

As you can see in Figure 13.7, the CET deals with measures depending on the time of implementation of the mitigations (M1, M2, etc.). It tries to establish effects regarding the containment of radioactive pollution during incidents and accidents in nuclear industry. This means that the probability of each measure being successful is not calculated.

In addition to this one (the containment of radioactive pollution), the CET can be developed for other consequences as different branches) if they are assumed to be critical during the incidents or accidents.

In this case, all of its branches should be considered because each of them presents a mitigation of a possible scenario and their outcome will have a

* You can also find it in literature as a Confinement Event Tree, but I think that there are big differences between these two event tree methods.

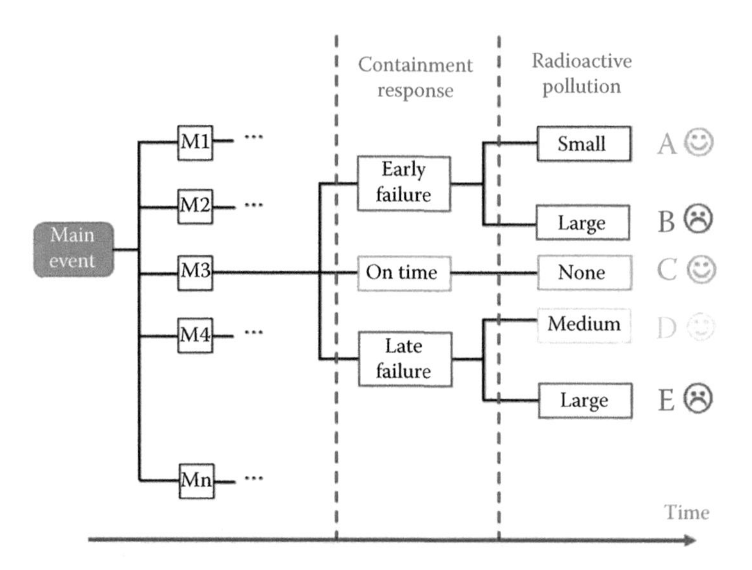

FIGURE 13.7
Containment event tree depending from time.

completely different effect on safety. Another difference from the conventional ETA* is that the consequences are presented on the top (horizontally with red letters on Figure 13.7) and the level of severity of outcomes is presented vertically.

The weaknesses here are that uncertainties regarding the outcomes are still present because these probabilities are not calculated. So, we do not know which of these outcomes is most common (has the highest probability of happening!). Also, the main context here is the failures and their effects, not their probabilities. Sometimes, the FTA is used for factors that are influencing the outcomes. In this case, the FTA helps us see the dependencies inside.

13.3.2 Dynamic Event Tree Analysis Method

Dynamic Event Tree Analysis Method (DETAM) is a method that is also used in nuclear industry. It takes care of the dynamics of how the scenario develops and is connected with the operator's activities. The nuclear industry as a high-risk industry is thoroughly regulated by many regulatory bodies. There are plenty of procedures that provide normal operations to be conducted smoothly. So, if anything abnormal happens (Main Event!), there are procedures on how to deal with these situations. These

* Conventional ETA is known also as Level 1 ETA.

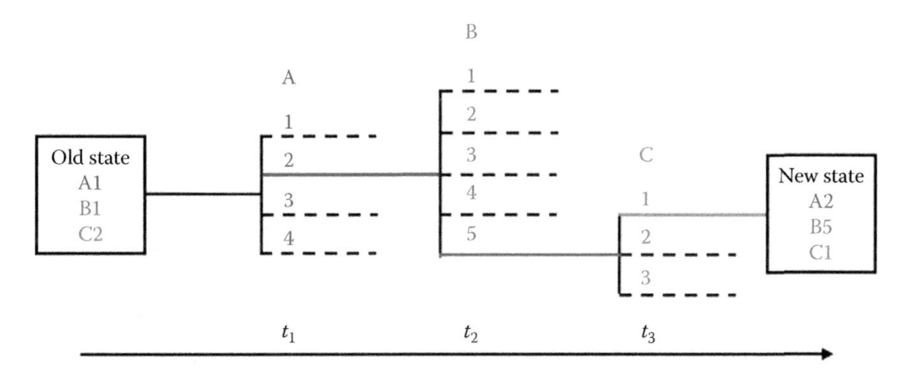

FIGURE 13.8
DETAM tree.

procedures are strongly dependent on the timely executions of particular steps by the operator in the Control Room. Figure 13.8 shows a simple example of the DETAM tree.

As you can notice in Figure 13.8, there is a diagram* in which particular actions need to be executed by the operator. These actions are part of an emergency procedure and are executed in their specified time. So, in a timely adjusted sequence, the operator must change the state of equipment from "old" to "new."

Actually, DETAM is mostly used for analyzing the operator's behavior (which is a strongly stochastic variable) in abnormal situations.

* Sometimes tables are used!

Final Words...

I do not hope that you have enjoyed this book!

This was not my intention: To make you to enjoy it!

This book is written with the intention to explain the Bowtie methodology as the best and the most holistic way to deal with quality or safety analysis in companies, especially those in risky industries. And it is my humble opinion!

I remember how much I struggled to get familiar with this methodology!

I hope that I have explained all aspects of the BM and provided enough reasons for you to learn and use it. This book can be used as a textbook for FTA and ETA separately, but its full potential can be materialized only if you use them together as the Bowtie methodology!

Please note that the latest edition of ISO 9001 published in 2015 is looking for *risk-based thinking* in maintaining quality. *Risk-based thinking* is something which is similar to pregnancy: A woman is either pregnant or nonpregnant, there is nothing in between! This means that there is no other way of implementing a *risk-based thinking* than to implement risk management. This is especially valuable for risky industries where the subjects must implement a safety management system (SMS) as requested by the regulatory bodies. Every company must implement two systems (SMS and QMS), which will be integrated in the future. In the future, I expect the integration of QMS and SMS due to their influence on each other, especially in the area of consequences. Another reason for this integration is the new development in safety presented by the concept of Safety-I and Safety-II. With this new approach, Safety-I is dealing with failures and Safety-II is dealing with successes. So, Safety-II is actually quality or better to say: Quality-I.

I could not find a book that dealt with the BM, so I decided to write one with the intention to provide quality learning material about the FTA and ETA (when used together for risk analysis, known as the Bowtie methodology). In the Internet, you can find a few articles and sites where the BM is mentioned, but not thoroughly explained. I remember when I was at a conference in the Delft University of Technology (the Netherlands) held in June 2015, a quality manager of a Japanese company mentioned that they are using both FTA and FMEA. On my question, why they needed both of them, he answered that it provides a holistic approach. I did not ask any further questions, but if he used the FTA and FMEA together for such a thing, he was obviously not very familiar with the BM...

I wish you good luck in your quality and safety analysis!

Index

Note: Page numbers followed by 'f' and 't' refer to figures and tables, respectively.

9781138067059